MW00364256

ROCKY MOUNTAIN ANIMAL TRACKS

Ian Sheldon & Tamara Hartson

LONE PINE

Distributed by: Canada Book Distributors - Booklogic
www.canadabookdistributors.com
www.lonepinepublishing.com
Tel: 1-800-661-9017

Canadian Cataloguing in Publication Data

Title: Rocky Mountain animal tracks / Ian Sheldon & Tamara Hartson.
Other titles: Animal tracks of the Rockies
Names: Sheldon, Ian, 1971– author, illustrator. | Hartson, Tamara, 1974–
   author.
Description: Illustrated by Ian Sheldon. | Previously published under title:
   Animal tracks of the Rockies. Edmonton : Lone Pine, 1997. | Includes
   bibliographical references and index.
Identifiers: Canadiana (print) 20200382829 | Canadiana (ebook)
   2020038290X | ISBN 9781774510292 (softcover) | ISBN 9781774510308
   (EPUB)
Subjects: LCSH: Animal tracks—Rocky Mountains—Identification. | LCSH:
   Animals—Rocky Mountains—Identification. | LCGFT: Field guides.
Classification: LCC QL768 .S53 2021 | DDC 591.47/9—dc23

Senior Editor: Nancy Foulds
Editors: Lee Craig, Nancy Foulds, Roland Lines
Production Manager: David Dodge
Design, layout and production: Gregory Brown
Technical review: Donald L. Pattie
Animal Illustrations: Gary Ross, Horst Krause
Track Illustrations: Ian Sheldon
Cover Illustration: Gary Ross
Scanning: Elite Lithographers Ltd., Edmonton, Alberta, Canada
Printing: Quality Color Press Inc., Edmonton, Alberta, Canada

We acknowledge the financial support of the Government of Canada.
Nous reconnaissons l'appui financier du gouvernement du Canada.

Funded by the Government of Canada | Canada
Financé par le gouvernement du Canada

PC: 38-1

Contents

Introduction

If you have ever spent time with an experienced tracker, or perhaps a veteran hunter, then you will realize just how much there is to learn about the subject, and just how exciting the challenge of tracking animals can be. Maybe you think that tracking is no fun, because all you get to see is the animal's prints. What about the animal itself; isn't that much more exciting? Well, for many of us who don't spend a great deal of time in the beautiful wilderness of the Rocky Mountains, the chances of us seeing the secretive Mountain Lion or the fun-loving River Otter are slim. The closest we may ever get to some animals will be through their tracks, and these can inspire a very intimate experience. Remember, you are following in the footsteps of the unseen—animals who are in pursuit of prey, or perhaps being pursued as prey.

This book offers an introduction into the complex world of tracking animals. Sometimes tracking is easy; other times it is an incredible challenge that leaves you wondering just what animal left those unusual tracks. Take this book into the field with you, and it can provide some help you with the first steps to identification. Prints and tracks are the book's focus; you will learn to recognize subtle differences for both. There are, of course, many additional signs to consider, such as scat and food caches, all of which help you to understand the animal you are tracking.

Remember, it takes many years to become an expert tracker. It is one of those skills that grows with you as you acquire new knowledge in new situations. Most importantly, you will have an intimate experience with nature. You will learn the secrets of the seldom seen. The more you discover, the more you will want to know, and by developing a good understanding of tracking you will gain an excellent appreciation of the intricacies and delights of our marvellous natural world.

How to Use This Book

First, take it into the field with you! Relying on your memory is not an adequate way to identify tracks. Tracking has to be done in the field, or with detailed sketches and notes. Much of the process of identification is circumstantial, so you will have much more success when standing beside the track.

The book is laid out to be easy to use. There is a quick reference appendix on p. 140 to the tracks of all the animals illustrated in the book. It is a fast way to familiarize yourself with certain tracks and the content of the book, and it guides you to the more informative descriptions of each animal and track. Each animal description is illustrated with appropriate footprints and styles of track that they frequently leave. These illustrations are not exhaustive, but show the most likely tracks or groups of prints that you will see. You will find a list of dimensions for the tracks, showing the general range, but there will always be extremes, just as there are with people who have unusually small or large feet.

If you think you may have identified a track but are not sure, then refer to the section on similar species. This is designed to speed up your consideration of which other animals leave the same or similar tracks, and should help firm up your conclusions.

Tips on Tracking

As you flip through this guide, you will notice clear, well-formed prints. Do not be deceived! It is a rare track that will ever show so clearly. For a good, clear print the perfect conditions will be slightly wet, shallow snow that isn't melting, or slightly soft mud that isn't actually wet. Needless to say, these events can be rare, and most often you will be dealing with incomplete prints, or faint ones where you cannot really be sure of the number of toes.

Should you find yourself looking at a clear print, then the job of identification is much easier for you. There are a number of key features to look for. Measure the length and width of the print, count the number of toes, check for claw marks and notice how far away they are from the body of the print, and look for a heel.

Keep in mind other more subtle features like the spacing between toes and whether they are parallel or not, and whether there is fur on the sole of the print making it less clear.

Do not rely on the measurements of one print, but collect measurements from several to give yourself an average impression. Even prints within one track can show a lot of variation.

When you are faced with the challenge of an unclear print—or even if you think you have made a successful identification from one print alone—it is time to think in the broader picture. Look beyond the single footprint and search out others.

Try to determine which is the fore print and which is the hind, and remember that many animals are built very differently from humans, having larger fore feet than hind feet. Sometimes the prints will overlap, or they can be directly on top of one another in a direct register. For some animals the fore and hind prints are pretty much the same.

Check out the pattern the prints make together in the track, and follow the trail for as many paces as is necessary for you to become familiar with the pattern. Patterns are very important, and can be the distinguishing feature between different animals.

Follow the trail for some distance, because it may give you some vital clues. For example, the trail might lead you to a tree, indicating that the animal is a climber, or down into a burrow. This can be the most rewarding part of tracking—you are following the life of the animal as it hunts, runs, walks, jumps, feeds or tries to escape a predator.

Take into consideration the habitat. Sometimes very similar species can only be distinguished by their exact location—one might be on the riverbank, while the other might be in the dense forest.

Think about your location too, and where you are within the mountains. Some animals have a limited range, which can rule out some species and help you with your identification. This book broadly categorizes distribution into the Canadian and American Rockies, with the latter further broken down into northern, central and southern regions.

Lastly, keep in mind that if you are quiet, you might catch up with the owner of the prints. Remember that every animal will at some point leave a print or track that looks just like a track of a different animal.

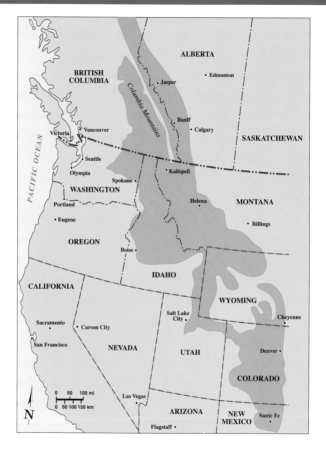

Terms and Measurements

Some of the terms used in tracking can be rather confusing, and they often depend on personal interpretation. For example, what comes to your mind if you see the word 'hopping'? Perhaps you see a person hopping about on one leg, or perhaps you see a rabbit hopping through the countryside. One person's perception of motion can be very different from another's. Below are some of these terms to clarify what is meant in this book, and where appropriate, how the measurement fits in with the term.

The following terms are sometimes used loosely and interchangeably. For example, while a rabbit might be described as a hopper, a squirrel seems to be more of a bounder, yet both leave the same pattern of prints in the same sequence.

Bounding: Can be used interchangeably with hopping, jumping; often used for short distances; hind prints usually registering ahead of fore prints.

Galloping: Used for the motion made by animals with four even legs, such as dogs, at speed, hind prints registering ahead of fore.

Hopping: Usually tight clusters of prints, with two fore prints set between and behind the hind prints; similar to bounding.

Running: Like galloping, but often applied generally to animals at speed.

Stotting: Confined to mule deer, describing the action of taking off from the ground at once, and landing on all four feet at once, in a pogo stick fashion.

Trotting: Faster than walking, slower than running.

Other Tracking Terms

Alternating track: As made in walk by humans, left-right sequence, often a double register for four-legged animals, which are described as diagonal walkers.

double register   direct register

Dewclaws: Two small toe-like structures set above and behind the main foot of most hoofed animals.

> dewclaws

dragline

Direct register: The hind print falls directly on the fore print.

Double register: The hind print falls slightly on or beside the fore print so that both can be seen at least in part.

Draglines: Lines left in snow or mud by foot or tail dragging over the surface.

Print: Fore and hind print treated individually; print dimensions measured are by length (including claws, maximum values may represent occasional heel register for some animals) and width; together prints make up a track.

Gallop group: Pattern made by prints in gallop, usually with hind prints registering ahead of fore prints, in a cluster of four.

Height: Taken at shoulder.

Length: Body length from head to rump, not including tail unless otherwise indicated.

Lope: Collection of four prints made at a fast pace, usually falling roughly in a line.

Retractable: Describing claws that can be pulled back in to keep them sharp, as with the cat family; claws do not register in print.

Sitzmark: Mark left on the ground by animal falling or jumping from a tree.

Straddle: Total width of the track, all prints considered.

Stride: Length from the center of one print to the center of the next print.

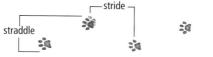

Track: Pattern left by a series of prints.

Trail: Often used to describe a track at length; think of it as the path of the animal.

For each species, the sizes of the fore and hind prints, stride and straddle are listed along with the weight and size of the animal. Under the category 'size' (of animal), the 'greater than' sign (>) is used when the size difference between sexes is pronounced.

As you read this book, you will notice an abundance of words such as *often, mostly* and *usually*. Unfortunately, it is very much the case that tracking will never be an exact science. We cannot expect animals to conform to our expectations, so be prepared for the unpredictable.

Mammals

Pronghorn Antelope

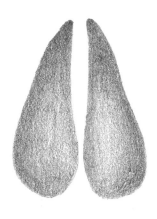

Fore and Hind Prints
Length: 3.3 in (8.4 cm)
Width: 2.5 in (6.4 cm)

Straddle
3.5–9 in (8.9–23 cm)

Stride
Walking: 8–19 in (20–48 cm)
Galloping: 14 ft (4.3 m)
    or more

Size (buck>doe)
Height: 3 ft (91 cm)
Length: 3.8–4.9 ft
    (1.2–1.5 m)

Weight
75–135 lb (34–61 kg)

walking

gallop group

Pronghorn Antelope

Antilocapra americana

This gracious antelope frequents the wide-open grasslands and sagebrush plains. Pronghorn are seldom seen in the Canadian foothills, but are much more abundant in the American Rockies. There, they gather in groups of up to a dozen animals in summer, and as many as 100 animals in winter. Unlike deer, Pronghorn run for fun, easily attaining constant speeds of 40 mph (64 km/h), or as much as 60 mph (97 km/h) for short bursts. The Pronghorn is one of the fastest animals in North America.

The Pronghorn print has a pointed tip and broad base. The hind prints usually register directly on top of the fore prints, making a tidy, alternating track. Pronghorn show a great tendency to drag their feet in snow, and they like to gather for feeding in areas where snow has been blown away. During their frequent gallops, the toe tips are spread wide, the dewclaws are absent and the distance between gallop groups increases the faster the antelope moves.

Similar Species: a deer print shows dewclaws, narrower toe bases and shorter strides between gallop groups; a Mule Deer (p. 28) frequents similar habitat, but its tracks show a stot pattern.

Bison

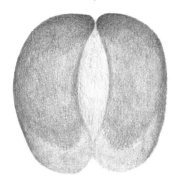

Fore and Hind Prints
Length: 4–6 in (10–15 cm)
Width: 4–6 in (10–15 cm)

Straddle
10–21 in (25–53 cm)

Stride
Walking: 14–32 in (36–81 cm)

Size (bull>cow)
Height: 5–6 ft (1.5–1.8 m)
Length: 10–12.5 ft (3–3.8 m)

Weight
Male: 800–2000 lb (360–900 kg)
Female: 700–1100 lb (315–495 kg)

walking

Bison
(Buffalo)
Bos bison

Bison, frequently called buffalo, once roamed North America in numbers estimated at 70 million. As few as 1500 remained after their slaughter during the 19th century, and thereafter a major effort began to save this magnificent beast from extinction. Today, as many as 100,000 Bison roam certain protected areas and public and private ranches, giving them a very scattered distribution. Do not be fooled by their calm exterior: 2000 pounds (900 kg) of bull Bison can inflict serious injury!

The hind print of the Bison is slightly smaller than the fore print, and usually registers on or near the fore print in an alternating walking pattern. In soft mud the whole print can be seen, but sometimes only the outer edge of the hoof registers on firmer ground. Foot drag is common, and the dewclaws may register in deep snow. Abundant pies may be mistaken for the droppings of regular cattle. For additional signs of Bison, look for rubbing posts or trees with tufts of distinctive, brown hair hanging from them, and for large pits dug into the earth for wallowing.

Similar Species: cattle; horses when Bison prints register on firm surfaces.

Mountain Goat

Fore and Hind Prints
Length: 2.5–3.5 in (6.4–8.9 cm)
Width: 2–3.3 in (5.1–8.4 cm)

Straddle
6.5–12 in (17–30 cm)

Stride
Walking: 10–19 in (25–48 cm)

Size (billy>nanny)
Height: 3–3.5 ft (91–107 cm)
Length: 4.8–5.9 ft (1.5–1.8 m)

Weight
100–300 lb (45–135 kg)

walking

Mountain Goat

Oreamnos americanus

Spotting the dazzling white coat of a Mountain Goat is truly a high mountain, wilderness experience. It is found in Canada and south into Idaho and Montana, and both reintroduction and introduction programs mean this goat is returning to its former territory and expanding its range. The Mountain Goat's keen vision and its remote setting high above the treeline make it difficult to see; its tracks in snow may be your best clue that the goat is around.

The goat's tracks show it has long, widely spreading toes, which produce a squarish print. The hard rim and soft middle of the foot help this agile goat clamber over the most unlikely crags at remarkable speeds, but even the Mountain Goat can make a fatal mistake! In deeper snow the feet may leave draglines, and the dewclaws may register. The alternating walking track is a double register, hind over fore. If you are having difficulty identifying prints, the goat's preference for a remote habitat may be your best indicator that it is around; few deer and sheep climb so high. In severe weather, the Mountain Goat may come down into deer territory.

Similar Species: deer (pp.28–31) and Bighorn Sheep (p. 24), except for their smaller, narrower and more pointed prints.

Bighorn Sheep

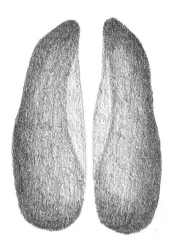

Fore and Hind Prints
Length: 2.5–3.5 in (6.4–8.9 cm)
Width: 1.8–2.5 in (4.6–6.4 cm)

Straddle
6–12 in (15–30 cm)

Stride
Walking: 14–24 in (36–61 cm)

Size (ram>ewe)
Height: 2.5–3.5 ft (76–107 cm)
Length: 4–6.5 ft (1.2–2 m)

Weight
75–275 lb (34–124 kg)

walking

Bighorn Sheep
(Mountain Sheep)

Ovis canadensis

In late fall, the loud crack of two majestic rams head-butting one another can be heard for a great distance. To watch the rut is an awe-inspiring experience. Widespread throughout the Rockies, this sheep prefers high, open meadows and scree slopes; it moves into valleys during winter. These sheep tend to avoid forested areas, and are not quite as bold as Mountain Goats on craggy cliffs.

The print is quite squarish in shape and pointed towards the front. The outer edge of the hoof is hard while the inner part is soft, giving the sheep a good grip on tricky terrain. Find one track, and you will likely come across several, because this sheep is gregarious and wants to be with others of its kind. The neat alternating walking pattern is a direct or double register of the hind over the fore. When running, the toes will spread wide. Their tracks may lead you to sheep beds—hollows dug into the snow, used many times and often with a large accumulation of droppings.

Similar Species: a deer print is more heart-shaped; a Mountain Goat's (p. 22) print is wider at the toe, and they rarely run; domestic sheep.

Mountain Caribou

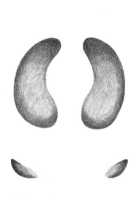

Fore and Hind Prints
Length: 3–4.8 in (7.6–12 cm),
   up to 8 in (20 cm)
   with dewclaws
Width: 4–5.8 in (10–15 cm)

Straddle
9–14 in (23–36 cm)

Stride
Walking: 16–32 in (41–81 cm)
Running: to 5 ft (1.5 m)
Group length: to 9 ft (2.7 m)

Size (buck>doe)
Height: 3.5–4 ft (1.1–1.2 m)
Length: 6.5–8.5 ft (2–2.6 m)

Weight
150–600 lb (68–270 kg)

walking
(on snow)

walking
(on hard ground)

Mountain Caribou

Rangifer tarandus

There are only a few places in the Rockies where you
might be privileged to see the elegant antlers of this true
wilderness animal, which is also called the Woodland Caribou. Its sensitivity to human encroachment has reduced
its range to the mountain parks in Canada, where it prefers to feed in groups above the treeline.

Distinctive, large, rounded prints spread wide to act like
snowshoes. Big dewclaws help caribou spread their weight
on snow, and tracks will reveal dewclaws registering well
behind the print. The faster caribou run, the more per-
pendicular the dewclaws are to the direction of travel.
Hind feet will rarely leave evidence of dewclaws. In snow,
foot drag is common, and from the foot drag you can
see how caribou swing their legs as they walk. In winter
the caribou's soft inner sole hardens and shrinks, leaving
a firm outer wall. Because of this outer wall, its hoofs leave
a neat circle on firmer surfaces. Look for scrapings where
caribou have tried to get to the lichen hidden underneath
the snow-cover.

Similar Species: by virtue of shape, size and range, there is
little chance for confusion.

Mule Deer

Fore and Hind Prints
Length: 2–3.3 in (5.1–8.4 cm)
Width: 1.6–2.5 in (4.1–6.4 cm)

Straddle
5–10 in (13–25 cm)

Stride
Walking: 10–24 in (25–61 cm)
Jumping: 9–19 ft (2.7–5.8 m)

Size (buck>doe)
Height: 3–3.5 ft (91–107 cm)
Length: 4–6.5 ft (1.2–2 m)

Weight
100–450 lb (45–203 kg)

walking   stot group

Mule Deer

Odocoileus hemionus

This wide-spread deer is found throughout the Rockies, and it is frequently seen in meadows, open woodlands and desert plains. In winter deer move down from higher terrain to warmer south-facing slopes and sagebrush flats, where they can still feed without having to contend with deep snow.

The Mule Deer has a neat alternating track with the hind print registering on the fore print. Deer frequently use the same well-worn trail in winter, preferring to stay in small groups. Their prints are heart-shaped and sharply pointed. In deeper snow or when deer are moving quickly, the dewclaws will register, and they are closer to the toes on the fore print. These deer have a unique gait for speed: they jump with all feet leaving and striking the ground at the same time. These stotting tracks will also show how the toes spread to distribute the weight and give better footing.

Similar Species: a White-tailed Deer (p. 30) print can be identical, but those animals prefer denser cover, and have a different gallop pattern with a shorter gallop stride; a Pronghorn (p. 18) has a wider base to the print; Elk (p. 32) and Moose (p. 34) have longer and wider prints.

White-tailed Deer

Fore and Hind Prints
Length: 2–3.5 in (5.1–8.9 cm)
Width: 1.6–2.5 in (4.1–6.4 cm)

Straddle
5–10 in (13–25 cm)

Stride
Walking: 10–20 in (25–51 cm)
Jumping: 6–15 ft (1.8–4.6 m)

Size (buck>doe)
Height: 3–3.5 ft (91–107 cm)
Length: up to 6.3 ft (1.9 m)

Weight
120–350 lb (54–158 kg)

walking   gallop group

White-tailed Deer

Odocoileus virginianus

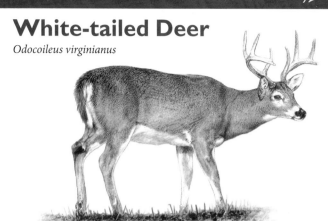

The keen eyesight of this deer guarantees that it knows about you before you know about it; frequently, all we see in the distance is its flashing white tail as it gallops away from us, which earns this deer its other name of 'flagtail.' These adaptable deer may be found in small groups at the edges of forests and in brushlands, and they are widespread throughout most of the Rockies. They can be common around ranches and residential areas.

The prints are heart-shaped and pointed, making an alternating track with direct or double registering hind prints on the fore prints. In snow or when the deer is galloping on soft surfaces, the dewclaws register. This flighty deer runs in the usual style, where hind prints fall in front of fore prints and the toes spread wide apart for the purpose of better footing.

Similar Species: readily confused with a Mule Deer (p. 28), but the habitat or gallop prints help in diagnosis; Elk (p. 32) and Moose (p. 34) prints are longer and wider; a Pronghorn (p. 18) prefers wide open spaces.

Elk

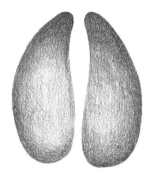

Fore and Hind Prints
Length: 3.2–5 in (8.1–13 cm)
Width: 2.5–4.5 in (6.4–11 cm)

Straddle
7–12 in (18–30 cm)

Stride
Walking: 16–34 in (41–86 cm)
Galloping: 3.3–7.8 ft (1–2.4 m)
Group length: to 6.3 ft (1.9 m)

Size (stag>hind)
Height: 4–5 ft (1.2–1.5 m)
Length: 6.5–10 ft (2–3 m)

Weight
500–1000 lb (225–450 kg)

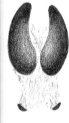

gallop print   walking

Elk (Wapiti)

Cervus elaphus

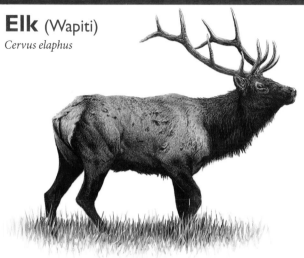

The widespread Elk can often be seen in social herds, feeding in forest openings and mountain meadows. The stag, however, prefers to go solo, and is easily recognized by his magnificent rack of antlers and distinctive bugling. Herds move into valleys when winter sets in.

Elk leave a neat alternating track with large, rounded prints, often in well-worn winter paths that they use frequently. Sometimes the hind print will double register slightly ahead of the fore print. In deeper snow, the dewclaws may register; this will also happen if the Elk gallops with its toes spread wide. A good place to look for Elk tracks is in the soft mud by summer ponds, where they like to drink and sometimes splash around.

Similar Species: most like a Moose (p. 34), but an Elk (p. 32) print is more rounded, has a narrower trail, shows foot drag more frequently in snow and is more often in groups.

Moose

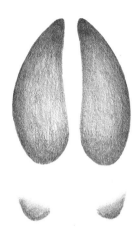

Fore and Hind Prints
Length: 4–7 in (10–18 cm),
   to 10.5 in (27 cm) with dewclaw
Width: 3.5–5.8 in (8.9–15 cm)

Straddle
8.5–20 in (22–51 cm)

Stride
Walking: 1.5–3 ft (46–91 cm)
Trot: to 4 ft (1.2 m)

Size (bull>cow)
Height: 5–6.5 ft (1.5–2 m)
Length: 7–8.5 ft (2.1–2.6 m)

Weight
600–1100 lb (270–495 kg)

trotting

Moose

Alces alces

The Moose, largest of all the deer, is impressive for its ungainly shape and massive rack of antlers. Solitary by nature, any groups of Moose are likely a cow with her calf. Despite their placid appearance, both the bull and cow will occasionally charge humans if approached.

Despite their shape, Moose are graceful movers, leaving a neat alternating track with the hind print directly or double registering on the fore. In summer look for tracks in mud by ponds and other wet areas where Moose are especially fond of feeding; Moose are excellent swimmers. In winter they feed in willow flats and coniferous forests, leaving a distinct browseline, or highline. Their long legs allow for easy movement in snow, and where the print is deeper than 1.2 inches (3 cm), the dewclaws will show far back from the main print, giving extra support to the animal's huge weight. Ripped stems and gnawed bark, 6 feet (1.8 m) or more above the ground, are additional signs that the Moose has been around.

Similar Species: an Elk (p. 32) has rounder, smaller prints with a narrower straddle and more foot drag, but is still readily confused with a juvenile Moose.

Horse

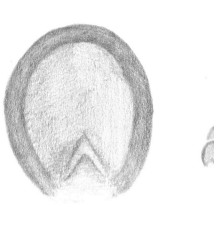

**Fore Print
(hind print is slightly smaller)**
Length: 4.5–6 in (11–15 cm)
Width: 4.5–5.5 in (11–14 cm)

Stride
Walking: 17–27 in (43–69 cm)

walking

Horse

Equus caballus

This popular animal has unmistakable prints, but deserves mentioning anyway as you will come across its tracks in many places. Back-country use of the horse means you can expect their tracks to show up almost anywhere. Unlike any other animals discussed in this book, the horse has one huge toe only, which leaves an oval print. A distinctive feature is the 'frog' or 'V'-shaped mark at the base. The horseshoe shows up clearly as a firm wall at the outside of the print. Not all horses will be shod, so don't expect to see this outer wall on every horse track. A typical, leisurely horse track is an alternating walk with hind prints registering on or behind the slightly larger fore prints. Horses are capable of a range of speeds—up to a full gallop—but most recreational horseback riders take a more leisurely outlook on life, preferring to walk their horses and soak up the mountain views!

Similar Species: a mule has smaller tracks and is rarely shod.

Coyote

fore

hind

**Fore Print
(hind print is slightly smaller)**
Length: 2.4–3.1 in (6.1–7.9 cm)
Width: 1.6–2.4 in (4.1–6.1 cm)

Straddle
4–7 in (10–18 cm)

Stride
Walking: 8–16 in (20–41 cm)
Gallop: 2.5–10 ft (76 cm–3 m)

**Size
(female is slightly smaller)**
Height: 23–26 in (58–66 cm)
Length: 32–40 in (81–102 cm)

Weight
20–50 lb (9–23 kg)

walking

gallop group

Coyote (Brush Wolf, Prairie Wolf)

Canis latrans

This widespread and adaptable canine prefers to hunt rodents and larger prey in open grassland or woodland; it hunts on its own, with a mate or as a family pack. A Coyote occasionally develops an interesting and cooperative relationship with a Badger, so you might just find their tracks together after they have been digging for ground squirrels.

The oval front prints are larger than the hind prints. Note the difference between the fore and hind heel pads. The hind heel pad rarely registers clearly. Claw marks are usually only evident on the two center toes. Coyotes typically walk or trot in an alternating pattern, the walk having a wider straddle. The trail left by a Coyote is often direct, as if it knew exactly where it was going, and in deep snow the tail hangs down, leaving a dragline. When galloping, the hind feet fall ahead of the fore feet, and the faster the speed, the straighter the gallop group.

Similar Species: a Domestic Dog's (*Canis familiaris*) print is not so oval and spreads more, the dog's four toes tend to register, and its trail is erratic and confused; a Red Fox usually has smaller prints, but there may be measurement overlap.

Gray Wolf

fore

hind

Fore Print
(hind print is slightly smaller)
Length: 4–5.5 in (10–14 cm)
Width: 2.5–5 in (6.4–13 cm)

Straddle
3–7 in (7.6–18 cm)

Stride
Walking: 15–32 in (38–81 cm)
Gallop: 3 ft (91 cm),
    leaps to 9 ft (2.7 m)

Size
(female is slightly smaller)
Height: 26–38 in (66–97 cm)
Length: 3.6–5.2 ft (1.1–1.6 m)

Weight
70–120 lb (32–54 kg)

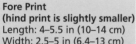
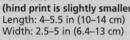

walking

trotting

Gray Wolf
(Timber Wolf)

Canis lupus

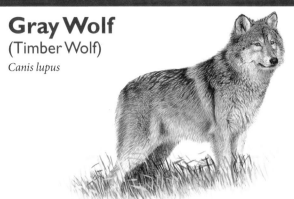

The soulful howl of the wolf epitomizes the outdoor experience, but it is not likely to be heard everywhere in the Rockies. While the wolf is more common to the north, reintroduction programs are gradually spreading its range southwards. The Gray Wolf, the largest of the wild dogs, works in packs in the wilderness and will rarely be seen.

A wolf leaves a straight alternating track with the smaller hind print registering directly on the front print. It is a large, oval print with all four claws showing. The lobing on the heel pads of the fore print are different from those on the rear print. If you find one track in the snow, it is likely the track of several, because wolves will sensibly follow their leader through deep snow, sometimes dragging their feet. When trotting, notice how the hind print has a slight lead and falls to one side, giving an unbalanced appearance. Wolves and Coyotes gallop in the same way.

Similar Species: a Domestic Dog's (*Canis familiaris*) print rarely can be as large, but the prints show a haphazard track where the register is not as direct, and the dog's inner toes tend to spread out more.

Gray Fox

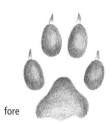

fore

hind

**Fore Print
(hind print is slightly smaller)**
Length: 1.3–2.1 in (3.3–5.3 cm)
Width: 1.1–1.5 in (2.8–3.8 cm)

Straddle
2–4 in (5.1–10 cm)

Stride
Walking/Trotting: 7–12 in (18–30 cm)

Size
Height: 14 in (36 cm)
Length: 21–29 in (53–74 cm)

Weight
7–15 lb (3.2–6.8 kg)

walking

Gray Fox

Urocyon cinereoargenteus

This small, shy fox is confined to the central and southern Rockies, preferring the woodland and brushland of the foothills. The front print shows a better register than the hind print, and the long, semi-retractable claws may not always register on the hind print. The heel pads are frequently unclear, sometimes just showing up as small, round dots. Tracks are in a neat alternating pattern for walking, with a trot like a Red Fox and a gallop group like a Coyote's. Follow the track, and you may be surprised to find this fox climbing a tree—the only fox to do so.

Similar Species: preferring the plains and deserts, but showing some range overlap, a Swift Fox (*Vulpes velox*) is a smaller fox with very similar, but frequently smaller, tracks and prints; the Red Fox (p. 44) is more widespread, has very different heel pads with a bar across, shows generally larger, less clear prints because of thick fur and has a longer stride and a narrower straddle, but measurements may overlap; a Gray Fox's trail is very much like a Domestic Cat's (*Felis catus*) or a small Bobcat's (p. 46), but its claws will show and the heel pad is smaller and more symmetrical.

Red Fox

fore

hind

Fore Print
(hind print is slightly smaller)
Length: 2.1–3 in (5.3–7.6 cm)
Width: 1.6–2.3 in (4.1–5.8 cm)

Straddle
2–3.5 in (5.1–8.9 cm)

Stride
Walking: 12–18 in (30–46 cm)
Trotting: 14–21 in (36–53 cm)

Size
(vixen is slightly smaller)
Height: 14 in (36 cm)
Length: 22–25 in (56–64 cm)

Weight
7–15 lb (3.2–6.8 kg)

walking   trotting

Red Fox

Vulpes vulpes

This beautiful and notoriously cunning fox is found throughout the Rockies in mountainous forests and open areas. It is very adaptable and intelligent. The fox leaves a distinctive, straight trail of alternating prints made up of a direct register of the hind print on the wider fore print. The feet are very hairy, obscuring the finer details of the footprints, and leaving only parts of the toes and heel pads showing. A very significant feature unique to this fox is the horizontal or slightly curved bar across the front heel pad. When trotting, the paired prints show the hind print falling to one side of the fore print in a typically canid fashion. The fox gallops in the same manner as the Coyote. The faster the gallop, the straighter the group of prints.

Similar Species: a Domestic Dog's (*Canis familiaris*) prints are of similar size, but they show no bar on the heel pad, and tracks show a shorter stride with a less direct trail; a small Coyote (p. 38), but it has a wider straddle; some overlap with a Gray Fox (p. 42), which has a shorter stride, wider straddle, and is found more often in the southern Rockies; a Domestic Cat (*Felis catus*) when its claws do not show.

Bobcat

fore

hind

Fore Print
(hind print is slightly smaller)
Length: 1.8–2.5 in (4.6–6.4 cm)
Width: 1.8–2.6 in (4.6–6.6 cm)

Straddle
4–7 in (10–18 cm)

Stride
Walking: 8–16 in (20–41 cm)
Running: 4–8 ft (1.2–2.4 m)

Size
(female is slightly smaller)
Height: 20–22 in (51–56 cm)
Length: 25–30 in (64–76 cm)

Weight
15–35 lb (6.8–16 kg)

walking   trotting to loping

Bobcat
(Wildcat)

Lynx rufus

The seldom-seen Bobcat is more widely distributed in the American Rockies, unlike the closely related Lynx, which is more dominant in the Canadian Rockies. As Bobcats are very adaptable animals, you will likely find tracks from wild mountainsides to residential areas.

The hind print is smaller than the fore; it usually registers exactly on the fore print in the walking pattern. Fore prints especially show asymmetry. Notice that the pad has two lobes to the anterior and three to the rear. As Bobcats pick up speed, their trail becomes a trot made of groups of two, hind leading the fore, and at even greater speeds the trail becomes a group of four prints in a lope pattern. Feet will leave draglines in several inches of snow. Unlike wild dogs, the trail meanders. Follow this trail far enough and you may find scrapings half burying scat—a sign of Bobcat marking its territory.

Similar Species: a Mountain Lion (p. 50) and juvenile Lynx (p. 48); prints of a large Domestic Cat (*Felis catus*) may be confused with a juvenile Bobcat's prints—a cat has a shorter stride, a narrower straddle and does not wander far from home, especially in winter; a dog's track shows claw marks, and the front of the footpad is lobed once only.

Canada Lynx

fore

hind

Fore Print
(hind print is slightly smaller)
Length: 3.5–4.5 in (8.9–11 cm)
Width: 3.5–4.8 in (8.9–12 cm)

Straddle
6–9 in (15–23 cm)

Stride
Walking: 12–28 in (30–71 cm)

Size
Length: 29–36 in (74–91 cm)

Weight
15–30 lb (6.8–14 kg)

walking

Canada Lynx

Lynx canadensis

This cat would be a thrill to see, but it eludes humans by living in dense forest. Because the Lynx is sensitive to human interference, it is abundant in the Canadian Rockies with more scattered distribution in the more populated American states. The Lynx, with its huge feet and relatively light body weight, stays on top of the snow while in pursuit of its main prey, the Snowshoe Hare.

This cautious walker leaves an alternating track with neat direct registers of the hind print over the fore print; it rarely drags its feet in deeper snow. Thick fur obscures the prints' characteristics, and they often appear as big, round depressions with no detail. The print may be extended in deeper snow by 'handles' off to the rear. However deep the snow, this cat sinks no more than 8 inches (20 cm). Its curious nature results in a meandering trail that may lead you to a partially buried cache of food.

Similar Species: a Bobcat's (p. 46) print is smaller; a Lynx is more likely to bound than run; a Mountain Lion (p. 50), which has clearer prints similar in size to a Bobcat's, sinks deeper because of its greater weight, and it has a wider straddle.

Mountain Lion

fore

hind

Fore Print
(hind print is slightly smaller)
Length: 3–4.3 in (7.6–11 cm)
Width: 3.3–4.8 in (8.4–12 cm)

Straddle
8–12 in (20–30 cm)

Stride
Walking: 13–32 in (33–81 cm)
Bounding: to 12 ft (3.7 m)

Size
Height: 26–31 in (66–79 cm)
Length: 3.5–5 ft (1.1–1.5 m)

Weight
70–200 lb (32–90 kg)

walking (fast)

Mountain Lion
(Cougar)
Felis concolor

The Mountain Lion is shy, elusive and nocturnal in nature, so finding this cat's tracks is usually the best trackers can hope for. Spread widely but sparsely because of its need for a big home territory, this large cat is essential in keeping the deer population down.

Prints tend to be wider than longer, and the retractable claws never register. In winter thick fur makes the track look much larger, and may obscure the two lobes on the front of the heel pad. When walking, the hind print will either directly register or double register on the larger fore print; as the speed of walking increases, the hind print will tend to fall ahead of the fore print. In snow the thick, long tail may leave a dragline, which can obscure some of the footprints' details. The Mountain Lion seldom gallops, but when it needs to, it is capable of quick bounding to catch prey.

Similar Species: a Bobcat's (p. 46) tracks may be confused with a juvenile Mountain Lion's; print size is similar to a Lynx's (p. 48), but a Lynx has narrower straddle, shorter stride, no tail drag and does not sink as deep in snow.

Fisher

fore left

Fore Print
Length: 2.1–3.9 in
(5.3–9.9 cm)
Width: 2.1–3.3 in
(5.3–8.4 cm)

Hind Print
Length: 2.1–3 in (5.3–7.6 cm)
Width: 2–3 in (5.1–7.6 cm)

Straddle
3–7 in (7.6–18 cm)

Stride
Walking: 7–14 in (18–36 cm)
Run: 1–4.2 ft (30–128 cm)

Size (male>female)
Length with tail: 34–41 in
(86–104 cm)

Weight
3–12 lb (1.4–5.4 kg)

walking   running

Fisher (Black Cat)

Martes pennanti

This agile hunter is comfortable both on the ground and in trees of mixed hardwood forests. Its speed and eager hunting antics make for exciting tracking— it races up trees and along the ground in its quest for squirrels. The Fisher's tracks may lead to the remains of an unfortunate Porcupine; it kills and eats them regularly, one of the few predators to do so.

Occasionally, the Fisher walks in a direct registering alternating track. However, it usually prefers to run in the typical weasel fashion of an angled pair of prints representing the direct register of hind over fore print. Five toes may register, but the small inner toe frequently does not. Only the front foot has a small heel pad that can show up in the print. Tracks often vary within a short distance, and are not associated with water: Fisher is a misnomer!

Similar Species: small female tracks may be confused with a male Marten's (p. 54), but a Fisher usually leaves much larger tracks; a Marten, which weighs less than a Fisher, will not leave such deep footprints; when the fifth toe doesn't register, a Fisher's print may be confused with a Bobcat's, so inspect the track pattern for typical mustelid habits.

Marten

right

Fore and Hind Prints
Length: 1.8–2.5 in
    (4.6–6.4 cm)
Width: 1.5–2.8 in
    (3.8–7.1 cm)

Straddle
2.5–4 in (6.4–10 cm)

Stride
Walking: 4–9 in (10–23 cm)
Running: 9–46 in (23–117 cm)

Size (male>female)
Length with tail: 21–27 in
    (53–69 cm)

Weight
1.5–2.8 lb (0.7–1.3 kg)

walking   running

Marten (Pine Marten)

Martes americana

This aggressive pred-
ator is found in
the coniferous mon-
tane and sub-alpine
forests. Size and habitat
are often key to identifying the Marten's
tracks. With many similarities to the Fisher
and Mink, the Marten can be mistaken for a small
version of the Fisher or a large version of the Mink.

In winter a very hairy foot often obscures all pad
detail from the print, especially detail from the
poorly developed palm pads. The heel pad is also very
undeveloped. Because of these features, the Marten
seldom leaves a clear track. Tracks often only show four
toes as the fifth inner toe fails to register. The hind print
registers on the fore in an alternating walking pattern.
The Marten bounds along in slightly angled groups
of two, hind falling on the fore print, in the pattern
common and typical of mustelids. Gallop tracks may
appear as clusters of three and four prints in a group.
Follow the criss-crossing tracks and you may find the
Marten has scrambled up a tree; look for the sitzmark
where it has jumped down again.

Similar Species: a large male Marten overlaps in size with
a female Fisher (p. 52), and a small female Marten overlaps with a male Mink (p. 60); Mink do not climb trees
and are often found near water unlike Martens; a Fisher
leaves clearer tracks.

Badger

fore

hind

walking

**Fore Print
(hind print is slightly shorter)**
Length: 2.5–3 in (6.4–7.6 cm)
Width: 2.3–2.8 in (5.8–7.1 cm)

Straddle
4–7 in (10–18 cm)

Stride
Walking: 6–12 in (15–30 cm)

Size
Length: 21–35 in (53–89 cm)

Weight
13–25 lb (5.9–11 kg)

Badger

Taxidea taxus

The squat shape and unmistakable face of this bold animal are features suitable for the open grasslands, although the Badger does also venture into high mountain country. Thick shoulders and forelegs, coupled with long claws, make for a powerful digging animal. Its long claws are evident in the pigeon-toed track that this member of the weasel family leaves as it waddles along.

Five toes register on both pairs of feet, although the claws on the hind feet are not as long. When walking, the alternating track is a double register, with the hind sometimes falling just behind the fore print and sometimes slightly ahead. Unlike other mustelids, the Badger likes to den up in a hole for the really cold months of winter, so look for the tracks in spring and fall snow. The Badger's wide, low body will often plough through deeper snow and obscure its own print detail.

Similar Species: a Badger's pigeon-toed tracks may resemble a Porcupine's (p. 104) in snow, but a Porcupine's tracks will show quill and tail draglines, and they will likely lead up a tree, not to a hole.

River Otter

fore left

hind left

Fore Print
Length: 2.5–3.5 in (6.4–8.9 cm)
Width: 2–3 in (5.1–7.6 cm)

Hind Print
Length: 3–4 in (7.6–10 cm)
Width: 2.3–3.3 in (5.8–8.4 cm)

Straddle
4–9 in (10–23 cm)

Stride
Walking/Running: 6–23 in (15–58 cm)

Size
(female is slightly smaller)
Length with tail: 3–4.3 ft (91–131 cm)

Weight
10–25 lb (4.5–11 kg)

running (fast)

River Otter

Lutra canadensis

No animal knows how to have more fun than the otter. If you are lucky enough to watch them at play, you will not soon forget the experience. Well-adapted for the aquatic environment, this otter is widespread along streams and other water bodies, so finding evidence of its presence is seldom a problem.

The five-toed feet leave evidence of webbing in soft mud, with inner toes set slightly apart. The front print does not show webbing so clearly. The heel may register, lengthening the print. Tracks are very variable, demonstrating the typical two-print bounding of mustelids, and with faster runs, groups of four and three prints. The thick, heavy tail often leaves a mark over the prints. In snow otters love to slide, often down riverbanks, leaving troughs nearly 12 inches (30 cm) wide. Otters play all year-round; in summer they roll and slide on grass and mud.

Similar Species: a Fisher (p. 52) has small tracks as does an otter, but a Fisher prefers forests, and does not leave conspicuous tail drag; when an otter enters a forest, it is usually going directly to another water body.

Mink

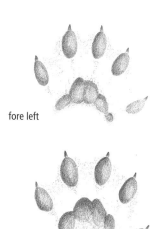

fore left

hind left

Fore and Hind Prints
Length: 1.3–2 in (3.3–5.1 cm)
Width: 1.3–1.8 in (3.3–4.6 cm)

Straddle
2.1–3.5 in (5.3–8.9 cm)

Stride
Walking/Running: 8–35 in (20–89 cm)

Size
(female is slightly smaller)
Length with tail: 19–28 in (48–71 cm)

Weight
1.5–3.5 lb (0.7–1.6 kg)

running

Mink

Mustela vison

The lustrous Mink prefers watery habitats surrounded by brush or forest, and it is widespread throughout the Rockies. At home as much on land as in water, this nocturnal hunter can be an exciting animal to track. Like the River Otter, the Mink slides in snow, leaving a trough carved out by its body for an observant tracker to spot.

The front print shows five toes, sometimes four, with five loosely connected palm pads in an arc. The heel pad on the front print rarely shows, but the hind heel may register on the hind print, which lengthens the print. The hind print only has four palm pads. The tracks show much diversity in gait, though the Mink prefers the typical mustelid bounding pattern of double prints, consistently spaced and slightly angled. Tracks may also appear as an alternating walk, or as a run with three and four print groups, as illustrated with the River Otter.

Similar Species: a large Mink may be confused with a small Marten (p. 54), but Martens stay away from water, and their tracks do not show such consistent double print bounding

Striped Skunk

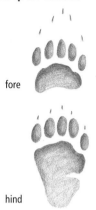

fore

hind

Fore Print
Length: 1.5–2.2 in
(3.8–5.6 cm)
Width: 1–1.5 in (2.5–3.8 cm)

Hind Print
Length: 1.5–2.5 in
(3.8–6.4 cm)
Width: 1–1.5 in (2.5–3.8 cm)

Straddle
2.8–4.5 in (7.1–11 cm)

Stride
Walking/Running: 2.5–8 in

(6.4–20 cm)

Size
Length with tail: 20–32 in
(51–81 cm)

Weight
6–14 lb (2.7–6.3 kg)

walking (fast)

running

Striped Skunk

Mephitis mephitis

   This striking skunk has a notorious reputation for its
vile smell. The Striped Skunk is widespread throughout
the Rockies, although less abundant in the most northerly
regions of Canada, and it prefers lower elevations and
diverse habitats. With such a potent smell for its defense,
the skunk rarely needs to run.
   Both feet have five toes, and long claws on the fore
print often show. Smooth palm pads and small heel pads
make for a print that is surprisingly small considering the
animal's size. Skunks mostly walk leaving a trail of prints
that rarely shows any consistent pattern, a distinctive
feature of this mustelid. An alternating walking pattern may
be evident, and the greater the speed, the more the hind
print oversteps the fore. Should the skunk need to run, the
track is a clumsy, four print pattern, groups set closely to
each other. In winter they den up, coming out on warmer
days and in spring. Their feet drag in snow. Often, the best
sign of their presence is their lingering odor drifting about.

Similar Species: a Western Spotted Skunk (p. 64) has
more southerly range and smaller prints in a very random
pattern.

Western Spotted Skunk

fore

hind

Fore Print
Length: 1–1.3 in (2.5–3.3 cm)
Width: 0.9–1.1 in (2.3–2.8 cm)

Hind Print
Length: 1.2–1.5 in (3–3.8 cm)
Width: 0.9–1.1 in (2.3–2.8 cm)

Straddle
2–3 in (5.1–7.6 cm)

Stride
Walking: 1.5–3 in (3.8–7.6 cm)
Jumping: 6–12 in (15–30 cm)

Size
Length: 13–25 in (33–64 cm)

Weight
0.6–2.2 lb (0.3–1 kg)

walking

running

Western Spotted Skunk
(Civet Cat)
Spilogale gracilis

This beautifully marked skunk is smaller than its striped cousin, and has a distribution through the American Rockies; it enjoys diverse habitats, such as scrubland, forests and farmland. The Western Spotted Skunk's nocturnal habits make it a rare sight; tracks are rare in winter as the skunk dens up to avoid the coldest months and only comes out on warmer days.

Well armed with a powerful smell, this skunk leaves a very haphazard trail as it forages for food on the ground, and occasionally climbs trees with ease. Long claws on the front foot will often register, and the palm and heel may show some defined pads. Although the skunk rarely runs, when it does it may bound along, leaving groups of four prints, hind ahead of fore. This skunk sprays only when very provoked, so the odor is less frequently detected than that of the Striped Skunk.

Similar Species: a Striped Skunk (p. 62) is bigger, with larger prints, less scattered tracks and a more widespread distribution to the north; a Striped Skunk does not climb trees, and it has a shorter running stride or it jumps.

Weasels

left

Least Weasel

Fore and Hind Prints
Length: 0.5–0.8 in (1.3–2 cm)
Width: 0.4–0.5 in (1–1.3 cm)

Straddle
0.8–1.5 in (2–3.8 cm)

Stride
Bounding: 5–20 in (13–51 cm)

Size (male>female)
Length with tail: 6.5–9 in (17–23 cm)

Weight
1.3–2.3 oz (37–65 g)

least weasel
(bounding)

Weasels

Mustela spp.

Long-tailed Weasel

Weasels are active hunters, with an avid appetite for rodents. Following their tracks can reveal much about these nimble creatures' activities. Active all year-round, the weasel's tracks are most evident in winter when the weasel will frequently burrow into the snow or, in pursuit of its prey, use an existing hole made by a rodent. Tracks may lead you up a tree from time to time, and weasels have been known to take to water.

The typical trail of the weasel is the bounding pattern of paired prints. Because of their light weight and small, hairy feet, pad detail is often obscured, especially in snow. Even with clear tracks, the fifth inner toe rarely registers. Successful identification of the weasel track can be troublesome. Your close attention to the straddle and stride is useful in tracking, but identification is further complicated by the small females of a larger species over-lapping with the large male of a smaller species. Clues can be gained from the habit displayed in jumping patterns. Distribution and habitat may also give hints about the identity of the mustelid's track.

Least Weasel

Mustela nivalis

The Least Weasel is the smallest of weasels, with the least well-defined track. This weasel has a scattered distribution in the Canadian Rockies only. Tracks may be found around wetlands, open woodlands and fields. Large males

Weasels

Short-tailed Weasel

Fore and Hind Prints
Length: 0.8–1.3 in (2–3.3 cm)
Width: 0.5–0.6 in (1.3–1.5 cm)

Straddle
1–2.1 in (2.5–5.3 cm)

Stride
Bounding: 9–35 in (23–89 cm)

Size (male>female)
Length with tail: 8–14 in
(20–36 cm)

Weight
1–6 oz (28–170 g)

Long-tailed Weasel

Fore and Hind Prints
Length: 1.1–1.8 in (2.8–4.6 cm)
Width: 0.8–1 in (2–2.5 cm)

Straddle
1.8–2.8 in (4.6–7.1 cm)

Stride
Bounding: 9.5–43 in
(24–109 cm)

Size (male>female)
Length with tail:
12–22 in (30–56 cm)

Weight
3–12 oz (85–340 g)

short-tailed weasel
(bounding)

long-tailed weasel
(bounding)

may leave the same sized tracks as Short-tailed Weasels, but Short-tailed Weasels do not frequent wet areas, preferring upland areas and woodlands.

Similar Species: a large male has measurement overlap with a small female Short-tailed Weasel (below).

Short-tailed Weasel (Ermine)
Mustela erminea

This weasel is larger than the Least Weasel, but smaller than the Long-tailed Weasel. It is widely distributed throughout the Rockies, preferring the woodlands and meadows up to higher elevations, and not favoring wetlands and the denser coniferous forest. Their bounding tracks may fall in clusters with a short stride, then a long stride repeated.

Similar Species: a small female has measurement overlap with a large male Least Weasel; a large male has measurement overlap with a small female Long-tailed Weasel (below).

Long-tailed Weasel
Mustela frenata

This weasel is the largest of the three, and shares some of the habitats of the Least Weasel, but is much more widely distributed throughout the Rockies. Their tracks—larger than the other weasels—show typical bounding with an irregularity in the length of stride, which is sometimes short and sometimes long, with no consistent behavior.

Similar Species: a small female has measurement overlap with a large male Short-tailed Weasel (above).

Wolverine

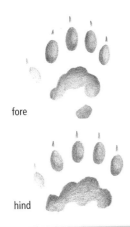

fore

hind

Fore Print
Length with heel: 4–7.5 in (10–19 cm)
Width: 4–5 in (10–13 cm)

Hind Print
Length: 3.5–4 in (8.9–10 cm)
Width: 4–5 in (10–13 cm)

Straddle
7–9 in (18–23 cm)

Stride
Walking: 3–12 in (7.6–30 cm)
Running: 10–40 in (25–102 cm)

Size
(female is slightly smaller)
Height: 16 in (41 cm)
Length: 32–46 in (81–117 cm)

Weight
18–47 lb (8.1–21 kg)

running (slow)

Wolverine

Gulo gulo

   This robust and powerful mustelid is the largest member of the weasel family, and its reputation has earned it many nicknames such as 'skunk bear' and 'Indian devil.' The Wolverine lives in the coniferous mountain forest, and its need for unadulterated wilderness has resulted in a scattered distribution in the central and southern American Rockies; more Wolverines are found to the north, especially in Canada.

   As with other members of the weasel family, there are five toes, and the fifth inner toe frequently does not register. The front print shows a small heel pad, but this pad rarely registers on the hind print. With its low, squat shape, the Wolverine leaves a host of erratic trails, all typical of mustelids. The track may be an alternating walking pattern, the typical bounding pairs of prints shown by all mustelids or the common track of groups of three and four prints in a lope.

Similar Species: when its inner toe does not register, its print may be confused with a wolf's, but its pad shape is very different; a Wolverine's tracks are erratic and much larger than other mustelids.

Raccoon

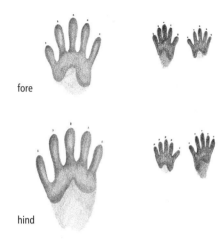

fore

hind

Fore Print
Length: 2–3 in (5.1–7.6 cm)
Width: 1.8–2.5 in (4.6–6.4 cm)

Hind Print
Length: 2.4–3.8 in (6.1–9.7 cm)
Width: 2–2.5 in (5.1–6.4 cm)

Straddle
3.3–6 in (8.4–15 cm)

Stride
Walking/Running: 7–20 in
    (18–51 cm)

Size (female is slightly smaller)
Length: 24–37 in (61–94 cm)

Weight
11–35 lb (5–16 kg)

walking   running group

Raccoon

Procyon lotor

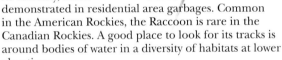

The inquisitive Raccoon is adored for its distinctive face mask, but considered a nuisance for its curiosity, often demonstrated in residential area garbages. Common in the American Rockies, the Raccoon is rare in the Canadian Rockies. A good place to look for its tracks is around bodies of water in a diversity of habitats at lower elevations.

The print looks like a human handprint, showing five well-formed toes. The small claws appear as dots. The highly dexterous forefeet rarely leave heel prints, but the hind prints, which are generally much clearer, do register a heel. The unusual walking track shows a front left next to a hind right that is then reversed on the next group. The fore print may be further ahead. Raccoons may also show an alternating track in deep snow, but this is rare because they den up in winter for the colder months. Occasionally, they run with hind prints falling ahead of fore prints in a cluster. Follow the trail and it may lead you up a tree, where Raccoons like to rest.

Similar Species: a Fisher (p. 52), which prefers deeper forests and shows different gaits; a Raccoon's (p. 72) track is like an Opossum's (p. 90) when unclear, but an Opossum drags its tail.

Ringtail

Fore and Hind Prints
Length: 1–1.4 in (2.5–3.6 cm)
Width: 1–1.4 in (2.5–3.6 cm)

Straddle
3–4 in (7.6–10 cm)

Stride
Walking: 3–6 in (7.6–15 cm)

Size
(female is slightly smaller)
Length: 24–32 in (61–81 cm)

Weight
1.5–2.5 lb (0.7–1.1 kg)

walking

Ringtail
(Cacomistle, Civet Cat, Miner's Cat)

Bassariscus astutus

This pretty cousin of the Raccoon is secretive and seldom seen, and is only found in the most southern mountain states. The Ringtail is strictly nocturnal, and rarely leaves any sign of its existence on the rocky terrain it frequents. When moving about, it will do so under the cover of shrubs, adding further to the difficulty of tracking this mammal. However, it will never be too far from water.

The small, rounded prints show five toes and the partially retractable claws will only register occasionally. A second pad is evident on the fore print, just behind the main pad, on a very rare basis. The common walking pattern is an alternating sequence of prints, where the hind registers on or close to the fore print. If you find a Ringtail's trail, it may lead you into rocky terrain, up a tree or to the animal's den.

Similar Species: a Domestic Cat (*Felis catus*) only has four toes, and does not have a second pad on the fore print; similar in size and shape to other small mustelids, but a Ringtail's gait and habitat are very different, and its fifth toe registers more than that of other mustelids.

Black Bear

right hind

right fore

Fore Print
Length: 4–6.3 in (10–16 cm)
Width: 3.8–5.5 in (9.7–14 cm)

Hind Print
Length: 6–7 in (15–18 cm)
Width: 3.5–5.5 in (8.9–14 cm)

Straddle
9–14.5 in (23–37 cm)

Stride
Walking: 17–23 in (43–58 cm)

Size (male>female)
Height: 3–3.5 ft (91–107 cm)
Length: 5–6 ft (1.5–1.8 m)

Weight
200–600 lb (90–270 kg)

walking
(slow)

Black Bear

Ursus americanus

The Black Bear is widespread in forested areas throughout the Rockies. Finding fresh bear tracks can be a thrill, as it means the bear may be just around the corner. It is always important to be cautious when in bear country; never underestimate the potential power of a surprised bear!

The bear's print is about the size of a human print, but wider and shows claw marks. The small inner toe rarely registers. The fore foot has a small heel pad that often shows, and the hind foot has a big heel. The bear's slow walk results in a slightly pigeon-toed double register of the hind over the fore. More frequently, at a faster pace, the hind oversteps the fore as shown with the Grizzly Bear. When running, the two hind feet register in front of the fore feet in an extended cluster. Along well-worn paths, look for digs—patches of dug-up earth—and bear trees, with their scratched bark and claw marks showing that these bears climb. Black Bears sleep deeply through winter so don't expect to find their tracks in the colder months.

Similar Species: a Grizzly's (p. 78) prints are often larger, its toes are set more in a line and closer together, and they have longer claws.

Grizzly Bear

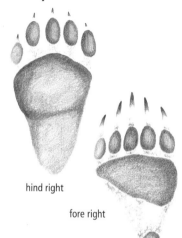

hind right

fore right

Fore Print
Length: 5–7 in (13–18 cm)
Width: 4–6 in (10–15 cm)

Hind Print
Length: 9–12 in (23–30 cm)
Width: 5–7 in (13–18 cm)

Straddle
10–20 in (25–51 cm)

Stride
Walking: 24–41 in (61–104 cm)

Size (male>female)
Height: 3–3.5 ft (91–107 cm)
Length: 6–7 ft (1.8–2.1 m)

Weight
240–1300 lb (110–600 kg)

walking (fast)

Grizzly Bear

Ursus arctos

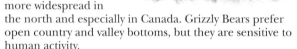

The magnificent and imposing Grizzly Bear symbolizes the Rocky Mountain wilderness, although it occurs elsewhere as well. This bear is very sparsely distributed in the American Rockies, but is more widespread in the north and especially in Canada. Grizzly Bears prefer open country and valley bottoms, but they are sensitive to human activity.

One of their huge prints will show four or five toes, with very long claws and a small heel pad on the fore print. A solid rear heel makes for a sturdy print. Toes are closely set in a line with the inner toe the smallest. The common walking track shows the hind registering ahead of the fore print, although a slower gait results in a trail like the Black Bear's. The Grizzly occasionally gallops. In winter the bear enters a deep slumber, meaning few of its tracks are seen. Well-worn trails may lead to digs, trees with claw marks high up on the trunk, or even the cache of a carcass. Take care if you find a cache, because the unpredictable Grizzly will likely be nearby.

Similar Species: a Black Bear (p. 76) has smaller prints, shorter claws and its toes are arranged more in an arc, and it makes territorial tree scratchings lower down a tree trunk.

Dusky Shrew

running group

Fore Print
Length: 0.2 in (0.5 cm)
Width: 0.2 in (0.5 cm)

Hind Print
Length: 0.6 in (1.5 cm)
Width: 0.3 in (0.8 cm)

Straddle
0.8–1.3 in (2–3.3 cm)

Stride
Running: 1.2–2 in (3–5.1 cm)

Size
Length with tail: 3–5 in (7.6–13 cm)

Weight
0.1–1 oz (3–28 g)

running

Dusky Shrew

Sorex monticolus

   While many species of tiny, frenetic shrews are found
in the Rocky Mountains, the most likely candidate is the
widespread and adaptable Dusky Shrew. This shrew is just
as happy in the high heathlands as in the lowland swamps,
but its rapid activity makes it difficult to observe closely.

   In its energetic and unending quest for food, the shrew
usually leaves a running pattern of four prints, but may
slow to an alternating walking pattern. In deeper snow
its tail often leaves a dragline. The individual prints in
a group are often indistinct, but more detail can be seen
on prints in shallow, wet snow or mud. In such conditions,
even the toes can be counted, with five on both fore and
hind feet. If you follow the shrew's trail it may disappear
down a burrow, and in snow its tunneling activity may leave
a ridge of snow on the surface.

Similar Species: there are many shrew species throughout
the region, including the Wandering Shrew (*Sorex
vagrans*), which prefers moist habitats, the larger
Water Shrew (*S. palustris*), often found by cold mountain
streams, and the more secretive and nocturnal Masked
Shrew (*S. cinereus*); mice show four toes on each of their
fore prints, unlike the Dusky Shrew.

Snowshoe Hare

hind

fore

Fore Print
Length: 2–3 in (5.1–7.6 cm)
Width: 1.5–2 in (3.8–5.1 cm)

Hind Print
Length: 4–6 in (10–15 cm)
Width: 2–3.5 in (5.1–8.9 cm)

Straddle
6–8 in (15–20 cm)

Stride
Hopping: 10 in–4.2 ft (25–128 cm)

Size
Length: 12–21 in (30–53 cm)

Weight
2–4 lb (0.9–1.8 kg)

hopping

Snowshoe Hare
(Varying Hare)

Lepus americanus

This hare is well known for its color change from summer brown to winter white, and its huge hind feet that enable it to 'float' on the surface of snow. Widespread, it frequents brushy areas in forests, as these areas provide good cover from the Lynx and Coyote, its most likely predators.

Like all rabbits and hares, its most common track is the hopping one, with groups of four prints in a triangular pattern. These groups can be long if the rabbit is running quickly. A hare track's most distinctive feature is that the hind print is much larger than the fore print. In winter heavy fur thickens the toes of the hind feet, and as the toes are separable they splay out, further spreading the hare's weight, when running. Follow the tracks and you might notice the well-worn runways that are often used as escape runs. If you are lucky, you might even come across a resting hare; hares are more active at night and do not live in burrows. For signs of their presence, look for twigs and stems that have been severed neatly at a 45 degree angle.

Similar Species: a Cottontail (p. 88) is smaller; a jackrabbit (p. 84) of open country is larger, but a Snowshoe Hare has the larger hind prints.

White-tailed Jackrabbit

hind

fore

Fore Print
Length: 2.5–4 in (6.4–10 cm)
Width: 1.3–1.7 in (3.3–4.3 cm)

Hind Print
Length: 3.5–6.5 in (8.9–17 cm)
Width: 1.5–2.5 in (3.8–6.4 cm)

Straddle
4.5–7 in (11–18 cm)

Stride
Hopping: 1–10 ft (30 cm–3 m)

Size
Length: 23–25 in (58–64 cm)

Weight
5–9 lb (2.3–4.1 kg)

hopping

White-tailed Jackrabbit

Lepus townsendii

This hare frequents the open country of the Rocky Mountain states, and is rarely seen in the Canadian foothills. These athletic animals can often be observed in the higher mountains.

Both fore and hind prints show four toes, and the hind foot may often register a long heel. The hopping action creates print groups in a triangular pattern, and these print groups spread out considerably as the hare speeds up. Living in the open country, this hare needs to be wary of predators. With its strong back legs, it is capable of leaping as much as 20 feet (6.1 m) to escape predators. Following the tracks could lead you to a depression where the hare rests, or they may reveal an urgent zigzag pattern, indicating that the hare had to flee from danger.

Similar Species: the measurements of a large Black-tailed Jackrabbit (*Lepus californicus*) may be the same as a small White-tailed Jackrabbit (p. 84), but the usually smaller Black-tailed is restricted to the southern Rockies; a Snowshoe Hare (p. 82) has much bigger hind prints, is not capable of such big leaps and requires dense cover; prints with no heel may look like a Coyote's (p. 38), but the gait is very different.

Pika

fore

hind

Fore Print
Length: 0.8 in (2 cm)
Width: 0.6 in (1.5 cm)

Hind Print
Length: 1–1.2 in (2.5–3 cm)
Width: 0.6–0.8 in (1.5–2 cm)

Straddle
2.5–3.5 in (6.4–8.9 cm)

Stride
Walking/Running: 4–10 in (10–25 cm)

Size
Length: 6.5–8.5 in (17–22 cm)

Weight
4–6 oz (113–170 g)

hopping

Pika (Cony, Rock Rabbit)

Ochotona princeps

When hiking high up in the mountains, you are more likely to hear the squeak of this cousin of the rabbit than see one. The widespread Pika rarely leaves good tracks because it lives on high mountain slopes in exposed rocky areas. Although active during winter, Pikas stay under the snow and feed on stored food. In summer they are quick to disappear under the rocks for protection.

Tracks may be found in spring when they fall on patches of snow or mud. The fore print shows five toes, although the fifth toe may not register, while the hind print has only four toes. Tracks may appear as an erratic alternating pattern or as three and four print bounding groups. The Pika's little hay piles, which dry in the sun in preparation for the long winter ahead, are more conspicuous signs of its presence.

Similar Species: a rabbit, although little confusion is necessary because of habitat; a Pika print may show five front toes, does not have a long heel and these prints are smaller and rounder.

Mountain Cottontail

hind

fore

Fore Print
Length: 1–1.5 in (2.5–3.8 cm)
Width: 0.8–1.3 in (2–3.3 cm)

Hind Print
Length: 3–3.5 in (7.6–8.9 cm)
Width: 1–1.5 in (2.5–3.8 cm)

Straddle
4–5 in (10–13 cm)

Stride
Hopping: 7–36 in (18–91 cm)

Size
Length: 12–17 in (30–43 cm)

Weight
1.3–3 lb (0.6–1.4 kg)

hopping

Mountain Cottontail
(Nuttall's Cottontail)

Sylvilagus nuttallii

This abundant rabbit is found widely through the American Rockies, but it is rarer to the north and in Canada. It prefers sagebrush and can be found in grassland and rockier areas.

As with other rabbits and hares, its most common track is the triangular grouping of four prints, with the larger hind prints falling ahead of the fore prints. The two fore prints might merge together. The hairy toes will obscure any detail you would otherwise hope for from the pads, and often the hind print can appear rather pointed. Following the tracks, you could be startled if the rabbit flies out from its 'form,' a depression in the snow or ground where it rests.

Similar Species: prints are like a Desert Cottontail's (*Sylvilagus auduboni*) or an Eastern Cottontail's (*S. floridanus*); a Desert Cottontail prefers dry lowlands in southern mountains, while an Eastern Cottontail is found in the eastern foothills; shares range with jackrabbits that leave much larger print clusters and longer strides; squirrel tracks show a similar pattern, but fore prints are more consistently side by side.

Virginia Opossum

fore

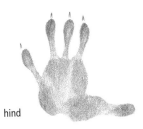

hind

Fore Print
Length: 2–2.3 in (5.1–5.8 cm)
Width: 2–2.3 in (5.1–5.8 cm)

Hind Print
Length: 2.5–3 in (6.4–7.6 cm)
Width: 2–3 in (5.1–7.6 cm)

Straddle
4–5 in (10–13 cm)

Stride
5–11 in (13–28 cm)

Size
Length: 2–2.5 ft (61–76 cm)

Weight
9–13 lb (4.1–5.9 kg)

walking

running

Virginia Opossum

Didelphis virginiana

This slow-moving marsupial is restricted to the southern American Rockies, but its range may be spreading. While found in many types of habitat, it still shows a preference for open woodland or brushland around water bodies, and is quite tolerant of residential areas. Tracks can often be seen in mud by water and in snow during the warmer months of winter—Opossums tend to den up during severe weather.

This rat-like animal is nocturnal and an excellent climber, so do not be surprised if its tracks lead to a tree. It has two walking habits: the common alternating pattern with hind print registering on the fore print, or the Raccoon-like pairs of prints with hind print next to opposing front print. The long, inward pointing thumb is very distinctive, and tracks show the claw is missing. In snow, the dragline from the long, naked tail may be stained with blood, because this unfortunate, thinly haired animal is not well adapted to the cold and frequently suffers from frostbite.

Similar Species: may be confused with a Raccoon (p. 72) if the print is unclear, as in sand or fine snow, otherwise its thumb is very distinctive.

Beaver

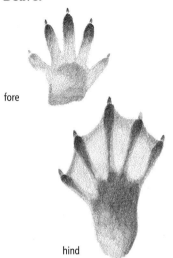

fore

hind

Fore Print
Length: 2.5–4 in (6.4–10 cm)
Width: 2–3.5 in (5.1–8.9 cm)

Hind Print
Length: 5–7 in (13–18 cm)
Width: 3.3–5.3 in (8.4–13 cm)

Straddle
6–11 in (15–28 cm)

Stride
Walking: 3–6.5 in (7.6–17 cm)

Size
Length with tail: 3–3.9 ft (91–119 cm)

Weight
28–75 lb (13–34 kg)

walking

Beaver

Castor canadensis

Few animals leave as many signs of their presence as the Beaver, the largest North American rodent and a common sight around water. Common signs of Beaver activity are the very evident dams and lodges—capable of changing the local landscape—and stumps of felled trees, with trunks gnawed clean of bark and marks from the Beaver's huge incisors.

The track of the Beaver is often obscured by its thick, scaly tail or by branches that Beavers drag about for construction and food. They often use the same path, resulting in a well-worn trail. If you do find tracks where individual prints are evident, look for webbing on the large hind prints. Broad nails, with the inner second toenail usually not showing, are another feature to look for in its hind print. Although the Beaver has five toes on each foot, it is rare for them all to register. The track may be in an alternating sequence with irregular prints, which results in a double register. Other signs of recent activity are scent mounds marked with castoreum, a yellowish fluid with a strong smell produced by the Beaver.

Similar Species: little confusion arises because a Beaver leaves many signs of its presence.

Deer Mouse

running group

Fore Print
Length: 0.3 in (0.8 cm)
Width: 0.3 in (0.8 cm)

Hind Print
Length: 0.6 in (1.5 cm)
Width: 0.4 in (1 cm)

Straddle
1.4–1.8 in (3.6–4.6 cm)

Stride
Running: 2–5 in (5.1–13 cm)

Size
Length with tail: 6–9 in
   (15–23 cm)

Weight
0.5–1.3 oz (14–37 g)

running

running
(in snow)

Deer Mouse

Peromyscus maniculatus

The most abundant mammal in the mountains is seldom seen because it is nocturnal. This highly adaptable rodent lives from arid valleys all the way up to alpine meadows, and may enter buildings to stay active during winter months.

Their fore prints each show four toes, three palm pads and two heel pads. The hind print has five toes with three palm pads; heel pads rarely register. It takes perfect, soft mud to register information from such a tiny mammal—it is a rare occurrence. Running tracks are most noticeable in snow, and they show two hind prints falling in front of two fore prints, which are closely set to each other. In soft snow the prints may merge and appear as larger pairs of prints with tail drag evident. Follow the tracks, and they may take you up the occasional tree or down into a burrow.

Similar Species: a Meadow Vole's (p. 98) merged two-print pattern has longer strides; a chipmunk has wider straddle; jumping mice may show measurement overlap; a House Mouse (*Mus musculus*) is very similar, but it is associated more with humans; a shrew (p. 80) has narrower straddle; identical to many less common species of mice.

Muskrat

fore

hind

Fore Print
Length: 1.1–1.5 in (2.8–3.8 cm)
Width: 1.1–1.5 in (2.8–3.8 cm)

Hind Print
Length: 1.6–3.2 in (4.1–8.1 cm)
Width: 1.5–2.1 in (3.8–5.3 cm)

Straddle
3–5 in (7.6–13 cm)

Stride
Walking: 3–5 in (7.6–13 cm)
Running: to 1 ft (30 cm)

Size
Length with tail: 16–25 in (41–64 cm)

Weight
2–4 lb (0.9–1.8 kg)

walking

Muskrat

Ondatra zibethicus

Like the Beaver, this rodent is found throughout the Rocky Mountains, wherever there is water. Beavers are very tolerant of Muskrats and even allow them to live in parts of their lodges. Muskrats are active all year, and they leave plenty of signs of their presence. They have an extensive network of burrows, often undermining the river bank, so if they have been very busy, do not be surprised if you suddenly fall into hidden holes! Small lodges out on water and beds of vegetation on which they rest, sun and feed during summer are other signs of this rodent.

The fore print has five toes, but one of these is reduced and rarely ever registers. The hind print has five well-formed toes. Together, they make up an alternating track with the hind print just behind or slightly over the top of the fore print. In snow the feet drag a lot, and the tail leaves a sweeping dragline.

Similar Species: few animals share this water-loving rodent's habits.

Meadow Vole

fore

hind

Fore Print
Length: 0.5 in (1.3 cm)
Width: 0.5 in (1.3 cm)

Hind Print
Length: 0.6 in (1.5 cm)
Width: 0.5–0.8 in (1.3–2 cm)

Straddle
1.3–2 in (3.3–5.1 cm)

Stride
Walking: 0.8 in (2 cm)
Running/Hopping:
    2–6 in (5.1–15cm)

Size
Length with tail:
    5.5–8 in (14–20 cm)

Weight
0.5–6 oz (14–170 g)

walking

running
(in snow)

Meadow Vole
(Field Mouse)

Microtus pennsylvanicus

A positive identification for a vole track is next to impossible, because, as with mice, there are many species to choose from. The Meadow Vole is commonly found in many damp or wet habitats—one possible way to differentiate this vole from other voles.

A vole print has four toes on the fore print and five on the hind print, although it is hard to tell; prints are seldom clear. The vole's walk leaves a paired alternating track, the hind print occasionally registering on the fore. The vole usually opts for the faster leap, where tracks appear in hopping pairs, hind registering on fore. Voles stay under the snow in winter, so look for distinctive piles of cut grass from their ground nests. The bark at the base of shrubs may show tiny teeth marks from gnawing. In summer voles use the same paths so often that they appear as little runways through grass.

Similar Species: a Montane Vole (*Microtus montanus*) is found commonly at higher elevations; a Southern Red-backed Vole (*Clethrionomys gapperi*) is also abundant; a Deer Mouse (p. 94) shows a paired hop pattern with shorter strides; a Northern Bog Lemming (*Synaptomus borealis*), which is restricted to the Canadian Rockies, has a four print diagonal loping pattern, not as common with voles.

Ord's Kangaroo Rat

slow hop group

**Hind Print
(fore print is much smaller)**
Length: 1.5–1.8 in (3.8–4.6 cm)
Width: 0.5–0.8 in (1.3–2 cm)

Straddle
1.3–2.3 in (3.3–5.8 cm)

Stride
Hopping: 5–24 in (13–61 cm)

Size
Length with tail: 8–13.5 in (20–34 cm)

Weight
1.5–2.5 oz (43–71 g)

fast hop

Ord's Kangaroo Rat

Dipodomys ordii

   This small, athletic rodent is capable of big jumps, as its name suggests. The best place to look for the nocturnal Ord's Kangaroo Rat is in the arid, shrubby regions of the southern American Rockies in sandy, dry soils. In cold weather they stay under the snow, venturing out on milder nights.

   Because of this rodent's preference for drier terrain, good tracks are hard to find, although sometimes an abundance of them can be found in sand, but the finer print detail will be lacking. The best way to identify the track is by its habit. When hopping slowly, two small fore feet register in between the large hind feet. These large hind feet will show a long heel mark. Also, look for the dragline left by the long tail. At speed, however, the fore prints are no longer evident, the hind heel appears shorter and the tail infrequently registers. If you find a good trail, it might lead you to the rodent's large nesting mounds. Find a burrow and tap your fingers by it. You may be surprised to hear something thumping back.

Similar Species: a Meadow Jumping Mouse prefers lush areas.

Bushy-tailed Woodrat

fore right

hind right

Fore Print
Length: 0.6–0.8 in (1.5–2 cm)
Width: 0.4–0.5 in (1–1.3 cm)

Hind Print
Length: 1–1.5 in (2.5–3.8 cm)
Width: 0.6–0.8 in (1.5–2 cm)

Straddle
2.3–2.7 in (5.8–6.9 cm)

Stride
Walking: 1.8–3 in (4.6–7.6 cm)
Jumping: 5–8 in (13–20 cm)

Size
Length with tail:
11–19 in (28–48 cm)

Weight
7–21 oz (198–595 g)

walking

running

Bushy-tailed Woodrat
(Packrat)

Neotoma cinerea

   This nocturnal woodrat is found almost everywhere, especially in rugged terrain and forests, but it is not fond of deserts, where other woodrat species reside. Tracking one of these rodents can be very rewarding as a trail might lead you to its distinctive mass of a nest, sometimes in an abandoned building; these nests can be 5 feet (1.5 m) across. This animal is a curious hoarder, bringing home all manner of objects, thus serving as a rather selective wilderness garbage collector.

   The woodrat often walks in an alternating fashion, where the hind print directly registers on the fore print; four toes register on the fore print and five on the hind. Its short claws will rarely show. This woodrat frequently runs, leaving a pattern of four prints with the larger hind print registering ahead of the diagonally placed fore prints. This animal's tracks tend to show short strides relative to the size of its feet. Do not be surprised if its trail ends at the base of a tree.

Similar Species: a Norway Rat (p. 108) is usually found close to human activity; a Marmot (p. 110) has similar but much larger prints; other less abundant species of woodrat.

Porcupine

fore

hind

walking

Fore Print
Length: 2.3–3.3 in (5.8–8.4 cm)
Width: 1.3–1.9 in (3.3–4.8 cm)

Hind Print
Length: 2.8–3.9 in (7.1–9.9 cm)
Width: 1.5–2 in (3.8–5.1 cm)

Straddle
5.5–9 in (14–23 cm)

Stride
Walking: 5–10 in (13–25 cm)

Size
Length with tail: 26–41 in (66–104 cm)

Weight
10–28 lb (4.5–13 kg)

Porcupine

Erethizon dorsatum

This notorious rodent has little need for running because its many long quills act as a formidable defense. Widespread throughout the Rocky Mountains, the Porcupine shows a preference for forests, but it can also be seen in more open areas.

As it has no need for speed, the most likely Porcupine track to be found is in an alternating walking pattern, with the longer hind print registering on or slightly ahead of the fore. Look for long claws on both prints. Note that the fore print only has four toes, not five as is often described. On clear prints, the unusual pebbly surface of its solid heel pads may show. It is evident from its track that this animal has a waddle for a gait, and its pigeon-toed footprints are often obscured by scratches from its heavy, spiny tail. In deeper snow, this squat animal drags its feet, and it may leave a trough with its body. The Porcupine's trail might lead you to a tree, where the animal spends much of its time feeding. Look for bark chewed from the trunk or nipped buds lying on the forest floor.

Similar Species: a Badger (p. 56) has pigeon-toed prints, no tail drag, and it doesn't climb trees.

Northern Pocket Gopher

fore

hind

walking

Fore Print
Length: 1 in (2.5 cm)
Width: 0.6 in (1.5 cm)

Hind Print
Length: 0.8–1 in (2–2.5 cm)
Width: 0.5 in (1.3 cm)

Straddle
1.5–2 in (3.8–5.1 cm)

Stride
Walking: 1.3–2 in (3.3–5.1 cm)

Size (male>female)
Length: with tail: 6–9 in (15–23 cm)

Weight
2.8–5 oz (79–142 g)

Northern Pocket Gopher

Thomomys talpoides

This seldom-seen rodent is found in all but the most northerly Rocky Mountains, from open pine forests to the high alpine meadows. It spends most of its time in burrows, only venturing out to move mud around and to find a mate. Because of their need for digging, pocket gophers prefer soft, moist soils.

By far, the best signs of Northern Pocket Gopher activity are the muddy mounds and tunnel cores that are especially evident after spring thaw. The mound will mark the entrance to the burrow, and it is always blocked up with a plug. Search around the mounds and you may find tracks. Both feet have five toes, and the front foot has well-developed, long claws for digging, although it is a rare track when you can make out this much detail. The typical track is an alternating walk, where the hind print registers on or slightly behind the fore print.

Similar Species: a Northern Pocket Gopher's tracks are associated with its distinctive burrows, which leave little room for confusion, except with other less abundant pocket gopher species.

Norway Rat

fore

hind

Fore Print
Length: 0.7–0.8 in (1.8–2 cm)
Width: 0.5 in (1.3 cm)

Hind Print
Length: 1–1.3 in (2.5–3.3 cm)
Width: 0.8–1 in (2–2.5 cm)

Straddle
3 in (7.6 cm)

Stride
Walking: 1.5–3.5 in (3.8–8.9 cm)
Jumping: 5–12 in (13–30 cm)

Size
Length with tail:
    12.5–18.5 in (32–47 cm)

Weight
7–18 oz (198–510 g)

walking

Norway Rat

Rattus norvegicus

   This despised rat is widespread almost anywhere in the American Rockies that humans have decided to build homes, although it is not entirely dependent on people and may live in the wild as well.

   Active both day and night, this colonial rat leaves tracks similar to the Woodrat's when it runs—groups of four with diagonally placed fore prints placed behind the hind prints. Commonly, the rat leaves an alternating walking pattern, where the larger hind print registers close to or on the fore, and the hind heel does not show. In snow the tail often leaves a dragline. The fore print shows four toes, while the hind has five toes. Since rats live in groups, you may find that there are many tracks locally, which often lead to their 2-inch (5.1 cm) wide burrows in the ground.

Similar Species: a Woodrat's (p. 102) tracks may be similar, although this animal rarely associates with human activity, except in abandoned buildings.

Hoary Marmot

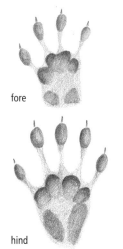

fore

hind

Fore and Hind Prints
Length: 1.8–2.8 in (4.6–7.1 cm)
Width: 1–2 in (2.5–5.1 cm)

Straddle
3.3–6 in (8.4–15 cm)

Stride
Walking: 2–6 in (5.1–15 cm)
Running: 6–14 in (15–36 cm)

Size (male>female)
Length with tail: 17–31 in
    (43–79 cm)

Weight
5–15 lb (2.3–6.8 kg)

walking

running

Hoary Marmot

Marmota caligata

These
endearing
squirrels
seem to
have a good life,
sleeping all winter
and sunbathing on rocks in summer. Hoary Marmots are
found in high rocky areas in the Canadian and northern
American Rockies. Marmots live in small colonies, with an
extensive network of burrows, and they are a joy to watch
when they play-fight.

The fore print has four toes with three palm pads and
two heel pads, although the heel pads are not always
evident. The hind print has five toes, four palm pads
and two poorly registering heel pads. When walking,
the marmot's hind print registers over its fore print
in the usual alternating walking pattern, but when running, the marmot will show a group of four prints, hind
ahead of fore. Because of marmots' preference for rocky
habitats, tracks are hard to come by, but they can be
found in spring and fall snowfalls.

Similar Species: a Yellow-bellied Marmot (*Marmota flavi-
ventris*) is found in the American Rockies, overlapping
the Hoary Marmot's range, and it likes similar habitats;
a Yellow-bellied Marmot is slightly smaller, but its measurements may be similar; a Hoary Marmot's running
tracks may be confused with a small Raccoon's (p. 72),
but a Raccoon has five toes on each fore print.

Red Squirrel

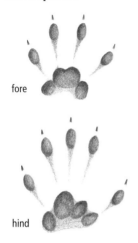

fore

hind

Fore Print
Length: 0.8–1.5 in (2–3.8 cm)
Width: 0.5–1 in (1.3–2.5 cm)

Hind Print
Length: 1.5–2.3 in
    (3.8–5.8 cm)
Width: 0.8–1.3 in (2–3.3 cm)

Straddle
3–4.5 in (7.6–11 cm)

Stride
Running: 8–30 in (20–76 cm)

Size
Length with tail: 9–15 in
    (23–38 cm)

Weight
2–9 oz (57–255 g)

running

running
(in deep snow)

Red Squirrel
(Pine Squirrel, Chickaree)

Tamiasciurus hudsonicus

You will realize when you have entered Red Squirrel territory because the squirrel greets you with a loud, chattering call. Other obvious signs of this widespread forest dweller are large middens—piles of cone scales and cores indicating a squirrel's favorite feeding site—left at the bottom of trees. Active year-round in their small territories, Red Squirrels leave an abundance of tracks from tree to tree or lots of tracks that may lead you down a burrow.

These energetic animals run mostly, in a gait that leaves groups of four prints, hind falling ahead of fore. The fore prints tend to be side by side, although this is not always the case. Fore prints each have four toes, with five on the hind, and the heels do not often register as squirrels run. In deeper snow its prints merge to form pairs of diamond-shaped tracks.

Similar Species: a Cottontail's (p. 88) fore prints rarely register side by side when it runs; a Chipmunk (p. 116) and Flying Squirrel's (p. 114) tracks show smaller straddle and smaller prints but similar pattern; in southern ponderosa pine forests, tracks could be from an Abert's Squirrel (*Sciurus aberti*), which has much larger dimensions in all respects.

Northern Flying Squirrel

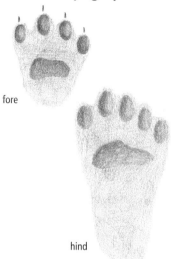

fore

hind

Fore Print
Length: 0.5–0.8 in (1.3–2 cm)
Width: 0.5 in (1.3 cm)

Hind Print
Length: 1.3–1.8 in (3.3–4.6 cm)
Width: 0.8 in (2 cm)

Straddle
3–3.8 in (7.6–9.7 cm)

Stride
Running: 11–29 in (28–74 cm)

Size
Length with tail: 9–11.5 in (23–29 cm)

Weight
4–6.5 oz (113–184 g)

sitzmark into running

Northern Flying Squirrel

Glaucomys sabrinus

These acro-
bats are found
in coniferous
forests at quite high
altitudes, in all but the southern-most American Rockies.
Their ideal habitat is widely spaced forest, where they
can make the most of the membranous flap between
their legs and glide through the night, from tree to
tree. Northern Flying Squirrels will den up together
in a tree cavity for warmth.

Because of its gliding, this squirrel does not leave as
many tracks as the Red Squirrel, and it is very difficult to
find any evidence of its presence in summer. In winter,
however, you might be lucky to come across a sitzmark.
Here, a squirrel lands on the ground leaving a distinctive
pattern (sitzmark), then rushes off, usually to the nearest
tree or to do some quick foraging. The bounding tracks
left in snow are typical of squirrels and other rodents,
with the hind prints falling ahead of the fore prints, and
the fore prints usually registering side by side.

Similar Species: a Red Squirrel's (p. 112) tracks are usually larger, and unlike a Northern Flying Squirrel's tracks,
they do not show a sitzmark, but in deep snow when
tracks are unclear, it can still be impossible to tell the
two squirrels apart; a Chipmunk (p. 116) has a smaller
straddle and smaller prints.

Least Chipmunk

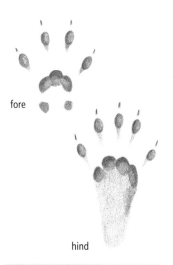

fore

hind

Fore Print
Length: 0.8–1 in (2–2.5 cm)
Width: 0.4–0.8 in (1–2 cm)

Hind Print
Length: 0.7–1.3 in (1.8–3.3 cm)
Width: 0.5–0.9 in (1.3–2.3 cm)

Straddle
2–3.1 in (5.1–7.9 cm)

Stride
Running: 7–15 in (18–38 cm)

Size
Length with tail: 7–9 in (18–23 cm)

Weight
1–2.5 oz (28–71 g)

running

Least Chipmunk

Tamias minimus

This delightful chipmunk is found in different habitats, from dry sagebrush to mountain forests, and it is bold enough to be a popular visitor in campgrounds. You are more likely to see or hear this rodent, which is highly active during summer months, before you would notice its tracks. In winter the chipmunks hibernate, but occasionally they venture out on milder days.

The chipmunks are so light that their tracks rarely show finer details. The front foot has four toes, while the hind foot has five. The two heel pads of the front foot often don't register, and they are completely lacking on the hind foot. These chipmunks run on their toes, so their heels rarely show at all. Their erratic tracks are made up of hind prints registering ahead of fore prints, as happens with many of their cousins. The chipmunk's trail often leads to extensive burrows. Look for piles of nutshells on rocks as further indication of a chipmunk's recent presence.

Similar Species: the larger Yellow-pine Chipmunk (*Tamias amoenus*) has scattered distribution from rocky montane habitats to the treeline in central and northern regions, including Canada; a Red Squirrel (p. 112) has larger tracks; mice have smaller dimensions; mid-winter tracks are more likely to belong to squirrels; a Least Chipmunk (p. 116) falls in the lower range of the track dimensions listed below, while a Yellow-pine is in the upper range.

Golden-mantled Ground Squirrel

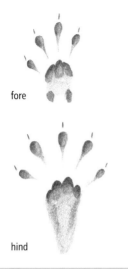

fore

hind

running

Fore Print
Length: 1–1.3 in (2.5–3.3 cm)
Width: 0.5–1 in (1.3–2.5 cm)

Hind Print
Length: 1.1–1.5 in (2.8–3.8 cm)
Width: 0.8–1.3 in (2–3.3 cm)

Straddle
2.3–4 in (5.8–10 cm)

Stride
Running: 7–20 in (18–51 cm)

Size
Length with tail: 8–13 in (20–33 cm)

Weight
6–10 oz (170–284 g)

Golden-mantled Ground Squirrel

Spermophilus lateralis

The Golden-mantled Squirrel is probably the most widespread of the several different ground squirrels in the Rocky Mountains, and enjoys different terrain, from open forests to rocky areas above the treeline. These bold squirrels can be very tame when they beg for food, but no matter how sweet they are, don't give in; our unhealthy diet certainly doesn't suit them!

As they hibernate during winter, their tracks may be evident in late or early snowfall or in mud around their many burrow entrances. The fore foot has such a reduced fifth toe that it rarely registers, and the two heel pads sometimes show. The larger hind foot has five toes. Both fore and hind feet have long claws that frequently are noticeable in tracks. Ground squirrels are usually seen scurrying around leaving a typical squirrel track—hind prints registering ahead of fore prints, which are usually placed diagonally.

Similar Species: a Richardson's Ground Squirrel (*Spermophilus richardsonii*) prefers open terrain in central and southern mountains; a Columbian Ground Squirrel (*S. columbianus*) and a Thirteen-lined Ground Squirrel (*S. tridecemlineatus*) are found in northern and central regions; a chipmunk is smaller; a tree squirrel's tracks have a more square-shaped running group.

Western Jumping Mouse

running group

running

Fore Print
Length: 0.3–0.5 in (0.8–1.3 cm)
Width: 0.3–0.5 in (0.8–1.3 cm)

Hind Print
Length: 0.5–1.3 in (1.3–3.3 cm)
Width: 0.5 in (1.3 cm)

Straddle
1.8–1.9 in (4.6–4.8 cm)

Stride
Hopping: 2–7 in (5.1–18 cm)
In alarm: 3–4 ft (91–122 cm)

Size
Length with tail: 7–9 in (18–23 cm)

Weight
0.6–1.3 oz (17–37 g)

Western Jumping Mouse

Zapus princeps

   Congratulations if you find and successfully identify the tracks of jumping mice! These mice are hard to come by, although they are found throughout the mountains, often in tall grass meadows. The preference of Western Jumping Mice for grassy meadows, and their habit of deep hibernation in winter make tracking very difficult. Their hibernation lasts about six months of the year!

   Their tracks are distinctive if you do find them. Clusters of four prints show, with the two smaller fore prints registering between the long hind feet. Their long heels do not always show. When jumping, these mice make short leaps, and the tail may show a dragline in soft mud or unseasonable snow. Clusters of cut grass stems, about 5 inches (13 cm) long and lying in meadows, are a more abundant sign of this rodent.

Similar Species: a Meadow Jumping Mouse (*Zapus hudsonius*) is found in the same habitats, but it is restricted to northern Canadian mountains in scattered populations; a Western Jumping Mouse's straddle overlaps that of a Deer Mouse (p. 94); a Kangaroo Rat (p. 100) also jumps, but on two hind feet, not all four.

Sasquatch

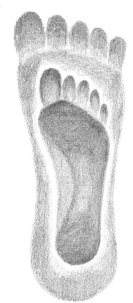

Foot Print
Length: average 14–17 in (36–43 cm)
Width: up to 7 in (18 cm)

Hand Print
Length: up to 12 in (30 cm)
Width: up to 7 in (18 cm)

Stride
Walking: 2–3 ft (61–91 cm)

Size (male>female)
Height: 6–8 ft (1.8–2.4 m)

Weight
400–1000 lb (180–450 kg)

Sasquatch

Imagine the thrill of being one of the few people to find footprints of the infamous Sasquatch! Bear prints can be impressively huge, and they are sometimes mistaken for Sasquatch's prints, but if you have found large tracks of undetermined origin, study them closely for differences. You will notice many distinctions between a bear's track and a track potentially made by a Sasquatch, such as prominent claws on bear prints.

This elusive, human-like inhabitant of the remote wilderness certainly has an enormous print. Its print is much broader and more flat-footed than our own—notice the raised arch on the human print. Most of Sasquatch's tracks have been found along rivers, but because Sasquatch could possibly weigh so much, a good, clear print can be seen on firmer surfaces. Tracks fall in a pattern just like our own when walking. Juvenile Sasquatch might be hard to distinguish from a large human footprint. If you think you have found a naked print of huge dimensions, be sure to let somebody know!

Note that a naked human footprint, which most of us should be familiar with, is best seen on sandy beaches by water. Here, humans can be seen lying around in large numbers, like lazy sea lions on the coast. Beyond the sandy beaches, however, the human print is usually shod to protect a person's sensitive soles from sticks and stones. Irresponsible humans leave plenty of evidence of their presence on trails. If the Woodrat hasn't beaten you to it, do the beautiful Rocky Mountains a big favor and pick up the garbage.

Birds & Amphibians

A guide to animal tracks is not complete without some consideration of the birds and amphibians found in the Rocky Mountains.

Birds come in all shapes and sizes; often there are very subtle differences between species that are not reflected in their tracks. The few birds chosen will demonstrate some of the major differences of birds common to the mountains.

Bird tracks can often be found in abundance in snow, and they are seen very clearly in shallow, wet snow. Of course, not all birds want to spend much time on the ground, perhaps preferring to be in the trees, or just soaring through the skies. Some species, on the other hand, spend a lot of time on the ground, and following their tracks can be entertaining. They can spin around in circles and lead you in all directions. Perhaps the track will suddenly end as the bird takes flight, or maybe it will end as a pile of feathers, the bird meeting its grizzly end at the claws, teeth or talons of a hungry predator. A most reliable source for bird tracks is the shores of streams and lakes, where mud can hold a print clearly for a long time. Many species of shore birds and waterfowl pick their way along through this mud, and the sheer number of tracks can be astonishing.

Many amphibians depend on moist environments; look in the soft mud along the shores of lakes and ponds for their distinctive tracks. Generally, frogs and toads have a different approach to getting around, so it is at least possible to tell these two types of tracks apart, remembering it is very difficult to differentiate between the individual species. In drier environments, most amphibians are ousted by reptiles since reptiles thrive in dryness; unfortunately, they seldom leave a good track for us to identify because of this preference for dry terrain.

Dark-eyed Junco

Dark-eyed Junco

Junco hyemalis

This small and common bird typifies the many small hopping birds found in the Rocky Mountain region. The total length of a Junco's footprint can be up to 1.5 inches (3.8 cm), with three forward pointing toes and one longer toe at the rear.

The exact dimensions of the toes may indicate what kind of bird you are tracking. Larger birds will have larger footprints—watch the footprint size when identifying these bird prints. Keep in mind what the season is too, as not all birds are present all year-round.

A good place to study these types of prints is by the birdfeeder. Watch the birds scurry around as they pick up fallen seeds, then have a look at the prints they have left behind. For example, Junco's are attracted to seeds that are scattered by chickadees when they forage for sunflower seeds in the birdfeeder.

The best prints are left in snow, although in deep snow the toe detail is lost, and the feet may show some drag-ging between the hops. Hop stride for these types of birds can be anywhere from 1.5 to 5 inches (3.8–12.7 cm), with a straddle of 1 to 1.5 inches (2.5–3.8 cm).

Common Raven

Common Raven

Corvus corax

This legendary bird spends a lot of time strutting around on the ground—confident behavior that perhaps hints at its intelligence.

The Raven shows a typical alternating track, with three thick toes pointing forward and one toe pointing back. The prints can be up to 4 inches (10.2 cm) long, with a stride of up to 6 inches (15.2 cm) and a straddle up to 4 inches (10.2 cm).

The track is similar to other corvids, all of which spend a lot of time poking around the ground: the Crow (*Corvus brachyrhyncos*) has prints up to 3 inches (7.6 cm) long, and the Magpie (*Pica pica*) has prints up to 2 inches (5.1 cm) long. Their strides are correspondingly shorter. When in need of greater speed, perhaps for takeoff, corvids leave a trail of diagonally placed pairs of prints that are rather irregular.

Ruffed Grouse

Ruffed Grouse

Bonasa umbellus

This ground-dweller prefers the quiet seclusion of coniferous forests in winter, so this will be the best place to find its tracks. Follow them carefully, and you might be startled when the grouse bursts from its cover underneath your feet. Its excellent camouflage usually affords it good protection.

The tracks of the three, thick, front toes are very clear, but the short one to the rear will not always show up so well. This toe is angled off to one side. The prints and straddle range from 2 to 3 inches (5.1–7.6 cm), with a stride of 3 to 6 inches (7.6–15.2 cm). Together, they make up a neat, straight track that perhaps reflects the bird's cautious approach to life on the forest floor. This grouse track is similar to others, such as the Blue Grouse (*Dendragapus obscurus*) and the White-tailed Ptarmigan (*Lagopus lecurus*), whose prints may be obscured by the winter feathers they grow on their feet, which make their feet appear larger.

Great Horned Owl

Great Horned Owl

Bubo virginianus

This wide-ranging owl is often seen resting quietly in trees during the day; it prefers to hunt at night. The mark left by this predator can be quite a sight if it registers well. You might stumble across this owl 'strike' and guess that the owl's target could have been a vole scurrying around underneath the snow. If you are a really lucky tracker, you will be following the surface track of an animal that abruptly ends with this strike mark—the animal has been attacked by an owl.

The owl strikes through the snow with its talons, leaving an untidy hole, which is occasionally surrounded by imprints of wing and tail feathers. These feather imprints are made as the owl struggles to take off with possibly heavy prey. The owl is an accomplished hunter in snow, although it is not the most graceful of walkers, and it prefers to fly away from the scene. Ravens can also leave this strike mark, but usually with much sharper feather imprints.

Canada Goose

Canada Goose

Branta canadensis

   This common goose is a familiar sight in open areas by lakes and ponds. Its huge webbed feet leave prints that can often be seen in abundance along the muddy shores of just about any water body, including in urban parks. Their droppings can accumulate in prolific amounts.

   The webbed foot has three long toes all facing forward. These toes register well, but the webbing between them does not always show on the print. The prints point inwards, giving the bird a pigeon-toed appearance, which perhaps accounts for the bird's waddling gait. A goose print measures 4 to 5 inches (10.2–12.7 cm) in length, with a 5 to 7 inch (12.7–17.8 cm) straddle and stride. These prints are typical of many waterfowl, most of which are smaller, such as the many species of gulls and ducks. If you come across exceptionally large prints similar in shape, they are likely from a swan.

Spotted Sandpiper

Spotted Sandpiper

Actitis macularia

The bobbing tail of this sandpiper is a common sight on the shores of lakes, rivers and streams. Teetering up and down on shores, they leave trails of three-toed prints. A fourth toe is very reduced and faces off to one side at an angle. While the sandpiper might be a common sight, you will usually only find one in any given place.

Their prints are about 0.8 to 1.3 inches (2–3.3 cm) long, and their tracks can have an erratic stride. These tracks are quite typical of all sandpipers and plovers, although there will be much diversity in size.

Frogs & Toads

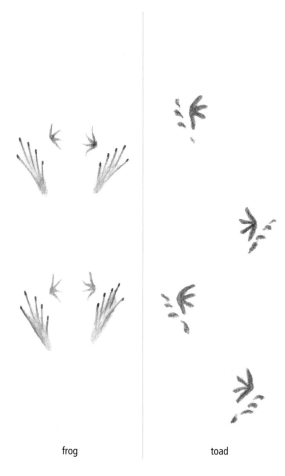

frog   toad

Frogs & Toads

The best place to look for frog and toad tracks is undoubtedly along the muddy fringes of water bodies, but toad tracks can occasionally be found in drier areas, for example as unclear trails in dusty patches of soil. Overall, frogs hop and toads walk. However, toads are pretty capable hoppers too, especially when being hassled by over-enthusiastic naturalists.

Among the frogs tracks that you might find is the Northern Leopard Frog (*Rana pipiens*), which can be found all the way up into moist mountain meadows. The hardy Spotted Frog (*R. pretiosa*) is widespread, but it frequents cold mountain streams and lakes, while the Wood Frog (*R. sylvatica*) is confined to the moist woodlands of northern regions. Hopping action is well displayed by all of

Western Toad

these, with two small fore prints registering in front of the long-toed hind prints. Such tracks may be up to 3 inches (7.6 cm) in straddle, depending on age and species.

The toad most likely to be found is the Western Toad (*Bufo boreas*), also called the Boreal Toad, which is found in streams, meadows and woodlands. This toad is widely distributed throughout the Rocky Mountains. Woodhouse's Toad (*B. woodhousii*) is found in moist lowland areas, sometimes even in temporary rain pools, in a few scattered pockets of the American Rockies. Toads leave rather abstract prints as they walk; the heels of the hind feet do not register. In less firm surfaces, you can often see the draglines left by the toes. The straddle may be up to 2.5 inches (6.4 cm).

Track Patterns & Prints

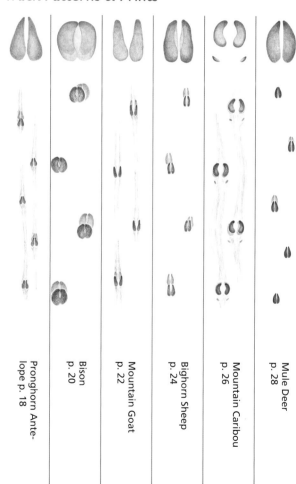

Pronghorn Ante-
lope p. 18

Bison
p. 20

Mountain Goat
p. 22

Bighorn Sheep
p. 24

Mountain Caribou
p. 26

Mule Deer
p. 28

Track Patterns & Prints

White-tailed Deer
p. 30

Elk
p. 32

Moose
p. 34

Horse
p. 36

Coyote
p. 38

Gray Wolf
p. 40

Track Patterns & Prints

Gray Fox
p. 42

Red Fox
p. 44

Bobcat
p. 46

Canada Lynx
p. 48

Mountain Lion
p. 50

Fisher
p. 52

Track Patterns & Prints

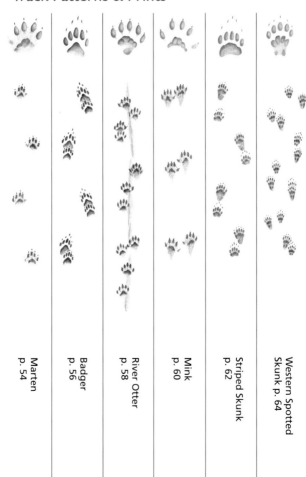

Marten
p. 54

Badger
p. 56

River Otter
p. 58

Mink
p. 60

Striped Skunk
p. 62

Western Spotted
Skunk p. 64

Track Patterns & Prints

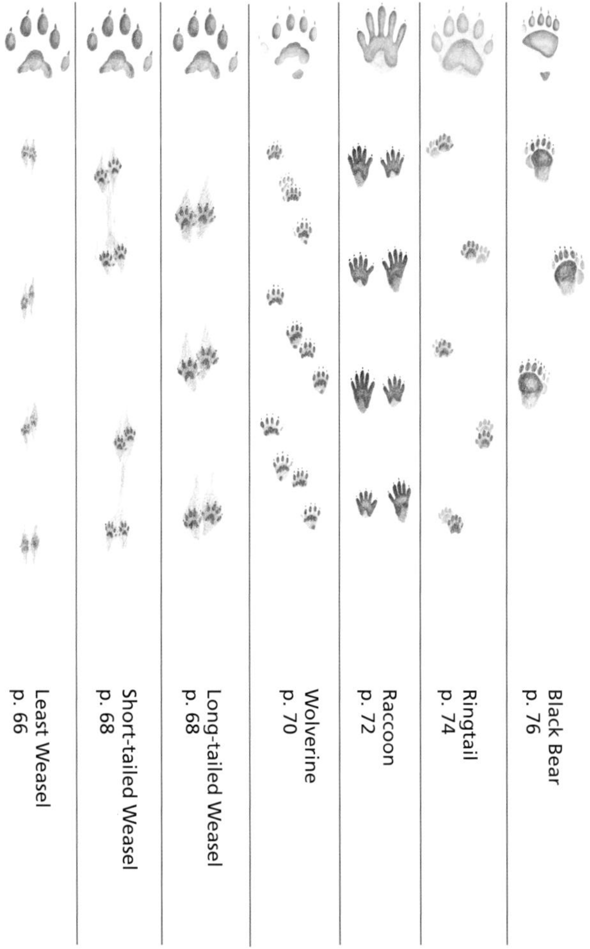

Least Weasel
p. 66

Short-tailed Weasel
p. 68

Long-tailed Weasel
p. 68

Wolverine
p. 70

Raccoon
p. 72

Ringtail
p. 74

Black Bear
p. 76

Track Patterns & Prints

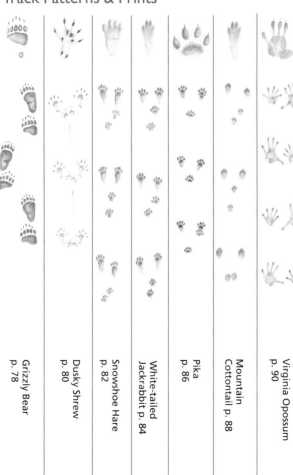

Grizzly Bear
p. 78

Dusky Shrew
p. 80

Snowshoe Hare
p. 82

White-tailed
Jackrabbit p. 84

Pika
p. 86

Mountain
Cottontail p. 88

Virginia Opossum
p. 90

Track Patterns & Prints

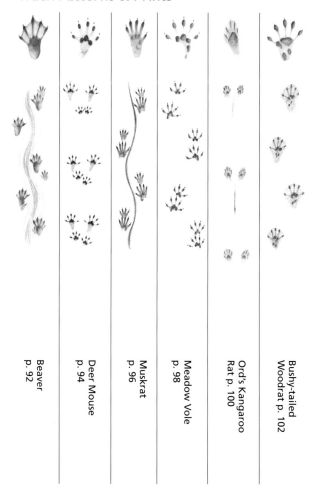

Beaver
p. 92

Deer Mouse
p. 94

Muskrat
p. 96

Meadow Vole
p. 98

Ord's Kangaroo
Rat p. 100

Bushy-tailed
Woodrat p. 102

Track Patterns & Prints

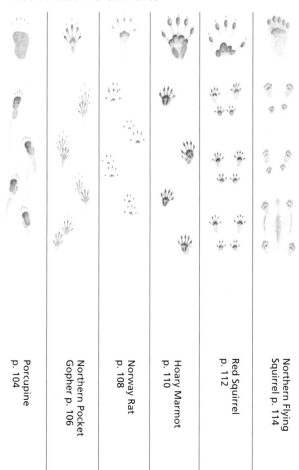

Porcupine
p. 104

Northern Pocket
Gopher p. 106

Norway Rat
p. 108

Hoary Marmot
p. 110

Red Squirrel
p. 112

Northern Flying
Squirrel p. 114

Track Patterns & Prints

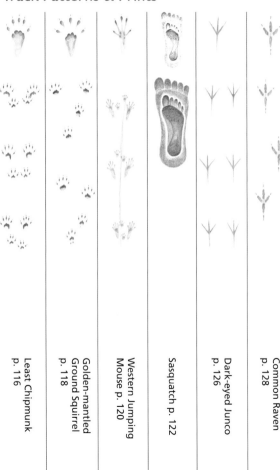

Least Chipmunk
p. 116

Golden-mantled
Ground Squirrel
p. 118

Western Jumping
Mouse p. 120

Sasquatch p. 122

Dark-eyed Junco
p. 126

Common Raven
p. 128

Track Patterns & Prints

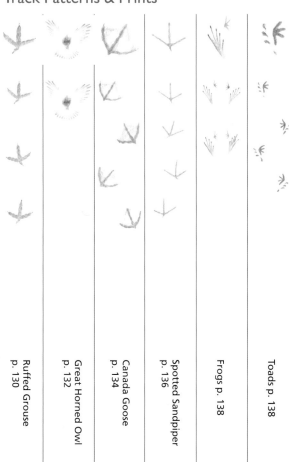

Ruffed Grouse
p. 130

Great Horned Owl
p. 132

Canada Goose
p. 134

Spotted Sandpiper
p. 136

Frogs p. 138

Toads p. 138

Print Comparisons

Fore Prints

Gray Fox

Red Fox

Bobcat

Coyote

Gray Wolf

Canada Lynx

Mountain Lion

Black Bear

Grizzly Bear

inch cm
0 — 0

1

2 — 5

Fore Prints

Long-tailed Weasel

Short-tailed Weasel

Least Weasel

Striped Skunk

Western
Spotted Skunk

Ringtail

River Otter

Marten

Mink

Wolverine

Fisher

Badger

inch cm
0   0

1

2   5

Hoofed Prints

White-tailed Deer

Mule Deer

Pronghorn Antelope

Bighorn Sheep

Mountain Goat

Elk

Horse

Bison

Mountain Caribou

Moose

inch cm
0   0
1
2   5

Hind Prints

Virginia Opossum

Muskrat

Hoary Marmot

White-tailed
Jackrabbit

Snowshoe Hare

Mountain Cottontail

Beaver

Porcupine

Raccoon

Hind Prints

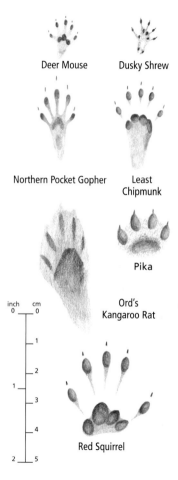

Deer Mouse

Dusky Shrew

Western
Jumping Mouse

Meadow Vole

Northern Pocket Gopher

Least
Chipmunk

Pika

Norway Rat

Bushy-tailed
Woodrat

Ord's
Kangaroo Rat

inch   cm
0   0

       1

       2

1   3

       4

2   5

Red Squirrel

Golden-mantled
Ground Squirrel

Northern Flying Squir

Bibliography

Barwise, J. E. 1989. *Animal Tracks of Western Canada.* Edmonton, Alberta: Lone Pine Publishing.

Behler, J. L., and F. W. King. 1979. *Field Guide to North American Reptiles and Amphibians.* National Audubon Society. New York: Alfred A. Knopf.

Burt, W. H. 1976. *A Field Guide to the Mammals.* Boston: Houghton Mifflin Company.

Farrand, J., Jr. 1995. *Familiar Animal Tracks of North America.* National Audubon Society Pocket Guide. New York: Alfred A. Knopf.

Forrest, L. R. 1988. *Field Guide to Tracking Animals in Snow.* Harrisburg: Stackpole Books.

Gadd, B. 1995. *Handbook of the Canadian Rockies.* Jasper, Alberta: Corax Press.

Halfpenny, J. 1986. *A Field Guide to Mammal Tracking in North America.* Boulder: Johnson Publishing Company.

Headstrom, R. 1971. *Identifying Animal Tracks.* Toronto: General Publishing Company.

Kavanagh, J. 1993. *Nature BC.* Edmonton, Alberta: Lone Pine Publishing.

Murie, O. J. 1974. *A Field Guide to Animal Tracks.* The Peterson Field Guide Series. Boston: Houghton Mifflin Company.

Rezendes, P. 1992. *Tracking and the Art of Seeing: How to Read Animal Tracks and Sign.* Vermont: Camden House Publishing.

Scotter, G. W., and T. J. Ulrich. 1995. *Mammals of the Canadian Rockies.* Saskatoon: Fifth House.

Stall, C. 1989. *Animal Tracks of the Rocky Mountains.* Seattle: The Mountaineers.

Stokes, D. and L. Stokes. 1986. *A Guide to Animal Tracking and Behaviour.* Toronto: Little, Brown and Company.

Wassink, J. L. 1993. *Mammals of the Central Rockies.* Missoula: Mountain Press Publishing Company.

Index

Page numbers in **boldface** type refer to the primary (illustrated) treatments of animal species and their tracks.

About the Authors

Ian Sheldon, an accomplished artist, naturalist and educator, has lived in South Africa, Singapore, Britain and Canada. Caught collecting caterpillars at the age of three, he has been exposed to the beauty and diversity of nature ever since. He was educated at Cambridge University and the University of Alberta. When he is not in the tropics working on conservation projects or immersing himself in our beautiful wilderness, he is sharing his love for nature. Ian enjoys communicating this passion through the visual arts and the written word.

Tamara Hartson, equipped from the age of six with a canoe, a dip net and a note pad, grew up with a fascination for nature and the diversity of life. She has a degree in environmental conservation sciences and has photographed and written about the biodiversity in Bermuda, the Galapagos Islands, the Amazon Basin, China, Tibet, Vietnam, Thailand and Malaysia.